PLEASURABLE KINGDOM

Animals and the nature of feeling good

Jonathan Balcombe

Macmillan
London New York Melbourne Hong Kong

First published 2006 by
Macmillan
Houndmills, Basingstoke, Hampshire RG21 6XS and
175 Fifth Avenue, New York, N. Y. 10010
Companies and representatives throughout the world

ISBN-13: 978–1–4039–8601–6
ISBN-10: 1–4039–8601–0

This book is printed on paper suitable for recycling and made from fully managed and sustained forest sources.

A catalogue record for this book is available from the British Library.

A catalog record for this book is available from the Library of Congress.

10 9 8 7 6 5 4 3 2 1
15 14 13 12 11 10 09 08 07 06

Printed and bound in China

to Maureen and Gerry

CONTENTS

ACKNOWLEDGMENTS

My agent, Sheila Ableman, was a tireless and loyal advocate from the start. Sara Abdulla was incredibly responsive and attentive for an editor with many projects to juggle.

This project benefited from so many in so many ways. I cannot thank them all, but I do especially want to thank the following: Marc Bekoff, Jaak Panksepp, Martin Stephens, Ken Shapiro, Gordon Burghardt, Russ Benford, Tom Regan, Mary Midgley, Michel Cabanac, Martin Harvey, Lorin Lindner, Masako Miyaji, Carolee Caffrey, Nathan Nobis, Manuel Marin, Nellie Tsipoura, Alan Sieradzki, Peter Marsden, Penny Patterson, Chris Sherwin, Jim Ackerman, David DeGrazia, Christi Richter Henson, Beth Barron, Stan Moore, Liz Day, Joyanne Hamilton, Alfred Rider, Sylvia Hope, Jo-Ann Jennier, Bob Falk, Ken Heyman, Walter Ratzat, Patti Finch, Kathy Gerbasi, Emily Patterson-Kane, Daniela Sharma, Rick Bogle, Helena Pedersen, Lynne Sneddon, Mary Thurston, Brock Fenton, Linda Case, Ellen Marsden, Reg Clark, and Joyce Poole. I thank Clark Wolf for the idea that the study of animal pleasure might be a new branch of ethology – what I term Hedonic Ethology (other suggestions welcome).

I am indebted to all the animals who taught me that quiet, patient observation rarely goes unrewarded. Finally, I thank Marilyn and Emily for being supportive as I, and they, navigated the pleasures and the inevitable pains.

Extracts from the following works are reproduced with the kind permission of their publishers:

Adams, R. *Watership Down.* Simon & Schuster.
Bagemihl, B. *Biological Exuberance*. Profile Books.
Burger, J. *The Parrot Who Owns Me.* Random House.
Burghardt, G. *The Genesis of Animal Play*. MIT Press.

Dawkins, R. *River Out of Eden.* Orion Books; Basic Books.
Dennett, D. *Kinds of Minds*. Perseus Group.
Goodall, J. and Bekoff, M. *The Ten Trusts*. Harper San Francisco.
Griffin, D. *Animal Minds*. University of Chicago Press.
Heinrich, B. *Mind of the Raven.* HarperCollins.
Linden, E. *The Octopus and the Orangutan.* Penguin USA.
Macphail, E. *The Evolution of Consciousness*. Oxford University Press.
Masson, J. and McCarthy S. *When Elephants Weep.* Random House.
Pollan, M. *The Botany of Desire*. Random House.
Steinhart, P. *The Company of Wolves*. Random House.
Tomkies, M. *Out of the Wild.* Random House.
Uhlenbroek, C. *Talking with Animals*. Hodder & Stoughton.
Wood Krutch, J. *The Great Chain of Life.* Morrow/HarperCollins.
Wright, R. *The Moral Animal.* Random House/Pantheon.
Young, R. *The Secret Life of Cows*. Farming Books and Videos.
de Waal, F. *Chimpanzee Politics*. Jonathan Cape.
de Waal, F. *Good Natured.* Harvard University Press.

Every effort has been made to trace the copyright holders of material reproduced herein; if any have been inadvertently overlooked the publishers will be pleased to make the necessary arrangement at the first opportunity.

Illustrations

Figures 1.2, 4.1, 6.1, 7.2, 8.1, 9.1, 9.2 and 11.2 reproduced with kind permission of Martin Harvey.

Figure 4.2 reproduced with kind permission of Jaak Panksepp and Oxford University Press.

Figure 5.1 reproduced with kind permission of Bat Conservation International.

Figure 8.2 reproduced with kind permission of Lorin Lindner.

PROLOGUE

Standing on the second floor balcony of a rustic hotel in Virginia during a spring birdwatching trip, I watched two fish crows land on an old wooden billboard protruding incongruously from a marshy landscape. I swiveled my telescope and focused on them. My perch was above theirs, and the angled morning sunlight was ideal for spying on these two black beauties.

First they pulled off some aerial antics, alternately launching and landing, sometimes trading places, each bird's attention riveted on the other. Then one bird sidled up to the other, leaned over and pointed her beak down, exposing her nape. The other bird gently swept his bill through her nape feathers as though searching for parasites. After a few seconds, the two edged apart again. Shortly, the one bird sidled back towards the other, and the grooming resumed. This process was repeated about 30 times over the next ten minutes. The groomed bird especially appeared to like it.

Male and female crows look the same to us, so I don't actually know whether these two were a mated pair or not. But whatever their sexes, their interaction appeared to embody the pleasure of contact.

So, why write a book on animal pleasure? Isn't it obvious that dogs love to chase balls and cats adore basking in the sun? Most wouldn't question that beloved companion animals enjoy these and other activities. But what of other animals we are less familiar with? What about warthogs, walruses, starlings, sparrows, iguanas, tree frogs, moray eels and puffer fish? Do they also experience pleasure? The answer I propose is undoubtedly yes, and the primary aim of this book is to present evidence, from both scientific study and anecdote, that the animal kingdom is rich with pleasure.

Demonstrating irrefutably that an animal is enjoying herself is not just difficult, it is practically impossible. While a dolphin or a dormouse may be behaving as if feeling joy, excitement, delight, or ecstasy, we can't be certain that she actually feels that way. And because we cannot know for certain how another animal feels, we have neglected, and in some cases censured, the study of emotions and feelings – including pleasure – in animals.

Happily, this situation is now changing. The academic study of animal minds and emotions is blossoming. Compassionate inquiry and reason are beginning to triumph over the narrow doctrine that has viewed animals as closed books. Sadly, the residue of past dogma remains. While the study of animal behavior – formally called 'ethology' – has grown into an important discipline, ethologists rarely entertain the idea of pleasure in animals. The word 'pleasure' appears rarely in the indexes of textbooks on animal behavior. Scientists openly discuss animal pain, which lies opposite pleasure on the spectrum of sensory experiences, and there are many books and journals dedicated to its study. Not so pleasure.

It is a stunning symptom of this scientific muteness that twenty-three centuries after Aristotle and 120 years post-Darwin there has not been one book dedicated to animal pleasure. This hole needs filling, and fast. As pleasure-seekers ourselves, we know that pleasure is not just the absence of pain, but a spectrum of positive feelings, felt in many ways and intensities. There are physical pleasures, like taste or touch, and there are psychological pleasures, such as music and art – even excitement, pride and relief.

Animal pleasure is an open frontier. I hope this book will prod researchers, ethicists, legislators, farmers, zookeepers, pet-owners and the rest of us to look for animal pleasure and to study it. History shows that we are much more likely to notice something when we look for it. If we think animals only have neutral or negative experiences (or none at all), then we are prone to overlook their positive moments. Researching and writing these pages has turned me on and tuned me in to the pleasures of animal existence. It is as though a curtain has been drawn open, allowing me to see nature in a new light. I now notice details that I missed before, even though they were there all along. I hope others will also become more attuned to the nuances of those 'other nations, caught with ourselves in the net of life and time, fellow prisoners of the splendor and travail of the earth,' to which writer and environmentalist Henry Beston so eloquently referred. May the world be richer for it.

A word on semantics. For simplicity and convenience, I use the term 'animal' to mean a creature other than the human animal. I don't mean to distance humans from other animals. On the contrary, I aim to illustrate the continuity of human sensory experience with that of other living things. In the same vein, I refer to a specific individual animal as a 'him' or a 'her,' and not the traditional 'it,' which reduces animals to mere objects. A table cannot feel pleasure; a tapir can. To treat animals as objects would undermine my belief that they are, like humans, unique individuals, whose lives are made better or worse by their circumstances.

Part I

WHY ANIMAL PLEASURE?

Chapter 1

SURVIVAL OF THE HAPPIEST

The adaptive basis for pleasure

When I hear a particular robin singing on a bough – I do not think: 'Irritable protoplasm so organized as to succeed in the struggle for existence.'

Joseph Wood Krutch, 1956

Science has neglected animal pleasure. Scientists like 'the big picture.' Research tends to focus on evolutionary explanations for natural phenomena. By considering only natural selection and reproductive success, it overlooks the experiences of individuals – their feelings, emotions, pleasures.

To appreciate the importance of pleasure to survival, consider the interplay of evolution and experience. Evolution concerns the adaptiveness of what an animal does (or doesn't do). It is the stuff of genes and survival. Experience, on the other hand, relates to an animal's conscious, sensory encounters with the world.

Evolution and experience are complementary, not exclusive. Just as an animal is the product of genetics and environment, so too do both evolution and experience guide decisions and behaviors. When an animal – let's say a raccoon – eats, she is satisfying a basic need of survival: to sustain herself. But in choosing, seeing, smelling and tasting food, she also experiences it. The physical pleasures of life – like the pains – are current, even though they have evolutionary significance. It is these experiences, not the evolutionary forces underlying them, that put wind in the sails of a raccoon's existence. And a mouse's. And a pigeon's.

Spicy foods and bathing ravens

Our own lives offer many ways to appreciate the evolution/experience dichotomy. If someone asks: 'What is the purpose of your life?,' you're unlikely to answer 'To maximize my reproductive output,' or 'I most want to ensure the propagation of my genes into the next generation.' You are more apt to say, 'I wish to lead a good life,' 'I seek to make the world a better place,' or 'I want to be happy and successful.' And by 'successful,' you are probably referring to how much you enjoy your life. As Aristotle had it: 'Life and pleasure ... are not separable; for without behavior there is no pleasure, and pleasure improves behavior.' Reproduction may be the currency of evolutionary success, but happiness is the currency of individual success. All that matters as far as evolution is concerned is that we survive long enough to reproduce. Individually, we seek a fulfilling, rewarding life, and plan to keep enjoying it long beyond our working years.

I doubt that many animals dwell on their 'careers'; they certainly don't set up retirement accounts. Nor are they likely to muse about maximizing their genetic output. They live – I suspect – mainly in the present, their actions motivated by desires, fears, instincts, and past experiences.

The use of spices in food provides a useful human illustration of the expeirential and evolutionary planes of existence. Scientists, in their quest for ultimate causation, have proposed several explanations for the evolutionary benefits of spicing food. Take your pick:

- spices provide micronutrients
- spices mask the bad taste of partially spoiled foods
- spices help cool us by making us sweat more
- spices combat harmful germs and microbes

Give yourself a point if you chose number four. Evidence currently supports the hypothesis that the antibacterial properties of spices account for our culinary habits, though any number of the above theories may work simultaneously.

Yet when you bite into a burrito are you thinking about banishing bacteria, mitigating malnutrition, dissipating distastefulness or, heaven forbid, swimming in sweat? No, you're enhancing the palatability of your food. Spices make food taste good. They promote pleasure. This is the proximate, experiential reason that we reach for the oregano or the curry powder. Our behavior may be beneficial, and it probably originates in our genes, but it is guided by our experience, by our senses, our desires and preferences. Pleasure rewards adaptive behavior. It is a vehicle by which nature promotes evolutionary success. Pleasure is one of the blessings of adaptation.

Figure 1.1 An agamid lizard enjoys the warmth of a perch.

A captive iguana stays on her perch beneath a warm sun-lamp rather than venturing into a colder region of her enclosure to retrieve a bit of food placed there. Shall we resort to some calculus of energetics and assume that the survival value of staying on the perch exceeds the survival value of getting the food? Or shall we just conclude that the iguana was so enjoying the warmth that it wasn't worth fetching the grub until hunger got the upper hand? University of Tennessee ethologist Gordon Burghardt, who made this observation, favors the latter interpretation, though he recognizes that staying put may be both the more pleasurable and the more adaptive decision on the part of the lizard (see Chapter 5 for a more hedonistic twist to the experiment.) Indeed, if natural selection favors behaviors that sustain life, then it should also favor rewards – such as pleasurable feelings – that reinforce those behaviors. Pleasure is adaptive, or as Canadian physiologist Michel Cabanac says it: 'pleasant is useful.'

Pleasure helps animals maintain a stable state. When we are cold, we seek warmth and it feels good. When we are hot, that same warmth no longer feels nice and we seek a cooler spot. When we're hungry, we eat and it tastes delicious. When we are full, we stop (usually). We know this from our experience, and from experiments. When human subjects are asked to dip their hand in a container of cool or cold water, they report the experience as pleasant if they are feeling hot (e.g. after emerging from a sauna), and unpleasant if they are feeling cold (e.g. after emerging from a freezer). The same phenomenon (Cabanac calls it 'alliesthesia' from the Greek: 'other-perception') applies to tastes (pleasant when hungry, unpleasant when full), though not to sounds and lights.

Alliesthesia applies to other animals. Nature rewards a cold animal who finds warmth, and vice versa. All an animal needs for alliesthesia to work is the capacity to experience surroundings as pleasant or unpleasant, and to move to a preferred environment. Sensory pleasure induces behaviors that improve homeostasis.

Here's another illustration of the dichotomy between evolution and experience. Many birds bathe – dipping their bodies, flapping their wings and shaking their feathers while standing in shallow water. American biologist Bernd Heinrich, working in the Maine woods, has compiled many observations of ravens bathing, and he acknowledges several possible ultimate, adaptive bases for this behavior. These include hygiene, combating skin parasites, and thermoregulation (maintaining a stable body temperature). But when a raven bathes, she is surely not aware of any evolutionary benefits. She is probably responding to a desire to get wet and cool, just as we enjoy the feeling of cool water or air on our skin when we are sweltering, or slipping into a hot tub on a wintry day.

So, while Darwinian evolution and survival undoubtedly influence animals' actions, animals aren't responding consciously to these influences. Yet it does seem that they are behaving according to their moods, their desires, and perhaps even a pre-planned daily schedule. Heinrich concludes that they do it simply because it feels good.

Pleasure (or reward) can be practical independent of its evolutionary origins. For instance, pleasure may enhance survival by reducing stress. Pleasurable activities stimulate the release of stress-reducing compounds into the body, such as opioids and endorphins. Prolonged stress and distress lower the body's defenses against disease in many animals, just as in humans. So by combating stress, the body's pleasure chemicals reduce vulnerability to any number of diseases and maladies.

A recent theory of feline purring is another variation on the theme of direct survival benefits from pleasure. The theory, proposed by Elizabeth von Muggenthaler of the Fauna Communication Research Institute in North Carolina, holds that a cat's purr has mechanical healing properties that speed up the repair of broken bones and other damaged tissues. Several cat species purr, including pumas, ocelots, servals, cheetahs and caracals, as well as the domestic cat. Purring may be cats' answer to ultrasound therapy in humans, which appears to improve bone growth and density. That purring in cats is believed to display contentment raises the question whether purring evolved first as a healing benefit, then later as a communication signal, or vice versa.

The idea that pleasure might be adaptive is not new. The prominent English animal behaviorist George Romanes in 1884 gave it the following Victorian spin:

> Pleasures and pains must have been evolved as the subjective accompaniment of processes which are respectively beneficial or injurious to the organism, and so evolved for the purpose or to the end that the organism should seek the one and shun the other.

Julian Huxley, in his treatise on animal language, also acknowledged the importance of individual experience, and it is a cornerstone of Charles Hartshorne's thesis that birds derive – not just deliver – a great deal of pleasure from their songs:

> Although the chief function of bird song is to maintain territory, it does not follow that the chief ... emotive meaning of singing for the bird is territorial hostility. ... Evolutionary

> causes of present behaviour lie deep in the past, but the animal is living now...

Another 20th-century ornithologist, Alexander Skutch, saw more than a struggle for survival in the lives of tanagers, flycatchers, elaenia and jacamars he studied in the New World tropics. He described the singing that accompanies nest building, incubation, and especially fledging of young in these species as 'a triumphant paean for the successful conclusion of their nesting.'

Are these the musings of starry-eyed scientists in moments of levity? Probably. Does that make them wrong? Probably not. The evolutionary importance of pleasure for motivating survival behaviors gives good reason to believe that they are right.

Some basic ingredients for pleasure

For pleasure to aid animals, they need the physical equipment to experience it. That we experience bliss, joy, comfort and satisfaction suggests that some other animals do too, because they are built like us in all the relevant ways. All vertebrates – mammals, birds, reptiles, amphibians and fishes – share the same fundamental body plan: a bony skeleton that supports a muscular system which enables the animal to move about, a nervous system that shuttles signals to different parts of the body and whose center of operations is the brain, a circulatory system that transports oxygen and other nutrients to body tissues, digestive and excretory systems that process food and eliminate wastes, a hormone system that helps regulate body processes, and a reproductive system evolved to ensure propagation.

To this shared foundation we can include a sensory system. All vertebrates have the same five basic senses as us: sight, smell, hearing, touch, and taste. There are some rare exceptions, such as the cave-dwelling fishes and salamanders that have gradually lost the capacity for vision, having lived for many generations in permanently dark environments. The senses are the interface between an animal's nervous system and its surroundings. The abilities to detect and avoid unpleasant stimuli and to seek rewards are the raw materials on which natural selection can act to favor pleasure and pain.

With all this equipment in common, it is no surprise that humans and animals share much of the same physiological and biochemical responses to sensory events. When we experience something painful or pleasurable, our brains signal our glands to secrete chemicals to help us deal with the situation. Human emotions are linked to two brain structures, the amygdala and the hypothalamus, and mediated by biochemicals including dopamine, serotonin and oxytocin. Many animals, especially mammals, possess these same neurological structures and brain chemicals as we do. That needn't necessarily mean they share our feelings, but careful observation of animals in action suggests that they do.

Recent brain imaging technologies such as PET and MRI provide further evidence that animals experience emotions somewhat like we do. There are, for example, remarkable similarities in brain regions in guinea pigs experiencing parent–offspring separation distress and in human brains during feelings of sadness. It is known that a variety of discrete emotional–behavioral control systems inhabit the same distinct regions of the brains of all mammals. According to American neuroscientist Jaak Panksepp, the core emotions – fear, rage, panic, play, seeking and lust – arise from the deep recesses of our primitive brains, and are believed to have evolved long before consciousness.

The brain releases dopamine in response to rewards like sex, food and water. The ability to produce dopamine has probably existed in animals for hundreds of millions of years. Even the humble sea pansy, a jellyfish relative, produces it, though probably not in relation to pleasure and pain. Goldfish prefer to swim in places where they have received amphetamine, which stimulates dopamine release from their brains. Such pleasure-promoting drugs as dopamine have probably played an important role in the evolution and maintenance of survival behaviors, at least in vertebrates.

Opiate receptors in human brains allow us to perceive pleasurable stimuli such as sweet tastes. Panksepp has shown that when rats play, their brains release large amounts of dopamine and opiates. When both humans and rats are given drugs that block these receptors, they rate the sweetness of normally 'liked' foods as less pleasant than normal.

Kent Berridge at the University of Michigan has devoted much of his career to the study of pleasure in the animal brain. Working mainly with taste in rats, Berridge's work suggests that brain networks cause 'liking' reactions to certain things in the animal's environment. These reactions suggest the conscious experience of pleasure (although in his view, pleasurable experience does not require consciousness). The crucial feature of positive states, he argues, is that potentially pleasurable events (e.g. the taste of sugar) be accompanied by positive patterns of behavior (e.g. licking of the lips).

Animals' brains appear to respond to many types of sensory pleasure, including food pleasure, drug pleasure and sex pleasure. More abstract forms of pleasure – including social joy, love, intellectual pleasures, aesthetic appreciation and even morality – are still largely unexplored, but as we shall see, interest in these is stirring.

A window on pleasure

Endowed with the brains, senses and biochemistry for responding to both rewarding and punishing stimuli, animals might be expected to behave accordingly. Because behavior is more readily observable than other aspects of a living animal (such as its biochemistry and physiology), it provides a practical – albeit imperfect – glimpse of their experiences. If other animals can experience pleasure, then we might reasonably expect them to behave in ways suggestive of pleasure in the sorts of situations that humans find pleasurable. The postures and expressions of domestic dogs and cats, for example, when receiving caresses from a trusted human, hint that they respond to the pleasure of touch something like we do – perhaps more so. That dogs and cats actively solicit more rubs or strokes bolsters the idea that they derive something beneficial from the stimulation.

A growing number of studies of wild animal populations, some spanning many years, even decades, are revealing nuances of animal lives that cursory studies miss. Chimpanzees, we now know, fashion tools and share food, and different populations have different cultural practices. African elephants communicate over distances of several kilometers using infrasound calls. Hyena societies are matriarchal and defend finely demarcated clan boundaries. Albatrosses mate for life and live for 70 years or more. Orca populations have language dialects. Honeyguides (a tropical bird) cooperate with honey-badgers and humans to get honey. And reef fish line up for 'cleaner' fishes to remove parasites and debris from their skin and inside their mouths and gills.

For many reasons, including their social intelligence and easygoing nature, rats are much studied, especially in the laboratory. They, too, show the behavioral hallmarks of pleasure. Studies by Panksepp and colleagues show that rats solicit tickles and strokes from trusted humans while making ultrasonic squeaks in the 50 kHz range (the upper limit of humans is around 20 kHz), just as they do when seeking sex and other rewarding social encounters such as rough-and-tumble play. Rats also become very active when placed alone in a Plexiglas chamber where they have been accustomed to playing with another rat. They vocalize and pace back and forth with excitement, in anticipation of play (see Chapter 4 for more about rat play).

Selfish genes and satisfied minks

When biologists seek to explain some aspect of an animal's behavior, physiology, or anatomy, they focus on adaptation for survival. We, too, are adapted for survival, and our behaviors are geared to it. Skeptics may scoff at the notion that we cultured humans spend much time engaging in survival activities. But we do. If you think that your nine-to-five job isn't a survival behavior, consider the prospect of stopping work. In the end, much as we might derive pleasure from it, we work for food and shelter just as other animals do – money merely being an intermediary resource with which to acquire these resources.

Similarly, survival ought to be rewarding for other animals. When a lioness 'goes to work' (e.g. hunts, cares for young) she isn't thinking about evolutionary success. Lions aren't taking time out to brush up on their knowledge of Darwinian fitness. Like us, they learn from experience, heed their desires and fears, and obey instincts. Survival and pleasure go paw in paw. Good feelings are powerful motivators of adaptive behaviors.

German has a word – *funktionslust* – for the pleasure and satisfaction one derives from doing what one is good at. More formally, *funktionslust* describes a theory that doing a behavior can enhance its motivation. Animals are naturally proficient at doing things important for their survival. Gibbons swing from branches, bats navigate using echoes, chameleons use their tongues like projectiles, and herons aim their dagger bills, all

Figure 1.2 A serval relishes her leaping prowess.

with striking skill. It has been said that four men with shovels can't keep pace with a burrowing aardvark. An impala can cross a paved road in a single leap, and a marlin can swim 100 meters in about four seconds.

Funktionslust is adaptive. Take the example of minks swimming. Swimming is important to survival in these highly aquatic members of the weasel family (who catch much of their food underwater). Better swimmers will tend to catch more food and, on average, reproduce more successfully, thereby passing along their genes for swimming prowess. If swimming is pleasurable for mink, then they are probably going to engage in it more often than they would if it weren't pleasurable, regardless of hunger. And because 'practice makes perfect,' mink who swim more will tend to be better at catching their aquatic food. Better-fed mink will tend to leave more offspring in the next generation, thereby conferring a selective advantage not only on swimming skill, but also on pleasurable feelings that motivate doing more of it.

A 2001 study of captive mink by Georgia Mason and colleagues at Oxford University found that they highly value access to a water pool, which enables them to satisfy their natural urge to dive and swim. The authors noted 'the key role of pleasure in motivating preference.'

Benefit of deterrence

As pleasure motivates preference, so too does pain motivate avoidance. Because pleasure, broadly speaking, is pain's counterpart on the continuum of sensory experience, animals' capacity for pain informs our understanding of their capacity for pleasure. Some of the physiological symptoms that accompany pain and distress mirror those that accompany certain sources of pleasure. Heart rate, blood pressure and breathing rate may all increase. If we accept that many animals feel pain, as most – including the majority of scientists – do, then we can also begin to explore whether most animals feel pleasure.

Like pleasure, pain is a catch-all term for a broad range of sensations. Just as pleasure encompasses such feelings as bliss, joy, anticipation, comfort, ecstasy, exhilaration, and satisfaction, we experience pain in a variety of forms and in different parts of the body. A headache feels different from a scrape, which differs from a burn, for instance. Some forms of pain and pleasure are less physical and more mental or emotional, like the pain of loss or the pleasure of pride. In the case of physical pain at least, if an animal has the equipment to experience it, then she is also probably equipped to experience pleasure.

The scientific study of pain is a thriving discipline. There are numerous journals dedicated to this subject. Many of these include studies of pain in animals.

Research over the past three decades increasingly suggests that many animals experience physical pain much as we do. Again, this should come as no surprise. We should expect all vertebrates – whose nervous systems are organized in fundamentally the same way as ours – to be alive to pain. Pain is valuable to animals which, having brains and legs (or fins) to move away from it, can learn to avoid and escape painful or threatening things. A painless animal, lacking the ability to associate pain with hazardous situations, would be less likely to contribute its genes for painlessness to future generations. Biologist Marian Stamp Dawkins puts it plainly:

> Pain evolved because, by being unpleasant, it keeps us away from the larger evolutionary disaster of death.

Besides helping us avoid the ultimate penalty of death, pain serves important functions, like keeping damaged body parts immobile while they heal.

Some scientists prefer the term 'nociception' to 'pain.' Unlike pain, nociception does not assume the animal can *experience* anything unpleasant, merely that its nervous system detects an aversive stimulus and reflexively recoils from it. However, when

Figure 1.3 The wait-a-bit thorn, one of many plants that exploit animal pain.

we show excessive reluctance to ascribe feelings to animals, we err grossly on the conservative side. Many papers appear in the pain journals describing the 'nociceptive' responses of mice to such standard pain-assessment methods as the immersion of the mouse's tail in hot water, or the commonly used tail-flick test, which uses radiant heat from a high-wattage bulb as a 'nociceptive' stimulus. Frogs are also used as a 'model' of mammalian pain. That science views these species as stand-ins for studying human pain conveys an implicit acknowledgment that they, too, can feel it. And while the experience of pain may differ in mice, frogs and humans, I find it more scientifically prudent to assume it hurts than that it doesn't.

The behavior of animals towards potentially painful stimuli shows that they can sense pain. Cats and dogs yowl and yelp when we inadvertently tread on their toes. Bee-eaters de-venomize bees by rubbing them against a twig and/or squeezing them to cause the venom to squirt out. When noxious substances are applied to the lips of trout, their heart rates increase, they rock side to side on their pectoral fins, and they take longer to resume feeding. Plants produce thorns, prickles, and harsh, bitter-tasting chemicals. Animals come armed with spines and horns. Among mammals alone, there are about 170 species that can make you feel like a pin-cushion, including hedgehogs, echidnas, tenrecs, porcupines, spiny mice, and spiny rats. Many fishes, lizards and even an amphibian use the same technique. Bizarrely, the sharp-ribbed newt, if grabbed by a predator, extends a row of sharp poison-tipped ribs through its own sides into the sensitive mouth lining of the hapless hunter. From the newt's perspective, having a sore midsection trumps being dead any day.

Animals, like humans, seek to relieve pain. Repeatedly, injured rats favor the bitter taste of water that contains a pain-relieving drug over unadulterated water. Fishes can learn to avoid noxious stimuli such as electric shocks and anglers' hooks. Chickens lame from poor husbandry ingest more food spiked with painkiller than healthy birds; and birds eat more drugged food the more lame they become.

None of these examples proves beyond all doubt that a vertebrate feels pain. Nor is it correct to assume that different species – or even different individuals – feel pain in the same way or at the same intensity. Sensory experience is complex and elusive – complex because it requires a whole, conscious organism to happen, and elusive because experiences are private. But the weight of evidence indicates that some animals – in particular those with backbones – can feel pain. Philosopher Steven Sapontzis of California State University reminds us that while our science likes to deny animal feelings, it implicitly acknowledges them. 'Our immediate experience of animals as feeling beings shows they have interests. The practice of performing pain and despair research on animals confirms this.'

Conclusion

If there is one most crucial reason why feeling good should not be the sole domain of *Homo sapiens*, it is this: pleasure is adaptive. Feeling good is a powerful motivator that steers animals towards behaviors that keep them alive and help them reproduce. Contrary to popular myth, life in the wild is not relentlessly harsh; survival and pleasure are mutually compatible.

Unsurprisingly, therefore, we find that animals have the equipment for experiencing pleasure as well as pain. Their brains and ours share common regions for processing sensory information, including rewards. The same sorts of chemicals course through their veins. And while most animals can't tell us in words how they are feeling (though some great apes have begun to), their behavior may convey much about what drives and deters them.

We are inextricably tied to other animals through our shared evolutionary origins. Our lives are buffeted by the same evolutionary pressures. All creatures – including humans – are trying to survive and pass their genes along to successive generations. But survival and reproductive success do not preoccupy animals' experience. Daily existence is colored by immediate experiences, not ultimate goals. The same is at least as true of other animals. Perhaps more so, for they probably live more in the moment than we do, as the Scottish poet Robert Burns said of a mouse:

Still, thou are blest, compar'd wi' me!
The present only toucheth thee:
But Och! I backward cast my e'e,
On prospects drear!
An' forward, tho' I canna see,
I guess an' fear!

Chapter 2

FORBIDDEN PLEASURE

Scientific taboos on animal feelings

There is a world of difference between what a scientist can publish and what we encounter in the world.

Eugene Linden, 2003

On a sunny but freezing cold January day in 2003 I stopped to watch starlings bathing at the edge of a row of small fountains outside the National Gallery of Art in Washington, DC. To me, starlings seem extroverts, rather like dogs.

About fifty birds were scattered in the trees nearby. Another dozen or so stood in the shallows, stooping down to dunk their heads, shake their wings, and dip their tails into the water. I followed the movements of individual birds, and noticed that they ventured into the water only for about a minute before flying back to the trees to preen, flutter their wings, and dry off in the sun. I also noticed that some birds were flying back to the fountains for repeated dips. They would land at the edge, wander in, dip and shake, wander out again, perhaps wander in once more, then fly purposefully back to the trees. Occasionally a human passerby would cause them all to fly at once, and a minute or two would pass before a brave bird would swoop down again, followed promptly by several more.

Were these birds enjoying their bathing and sunning? Were they feeling some measure of excitement or exhilaration as they immersed their bodies in the bracing water? Or were they performing some act vital – or at least beneficial – to their survival? Scientists are more comfortable making the latter, adaptive, sort of interpretations. The cold water may help dislodge fleas and mites from their skin, or perhaps clean away the dust and grime that accumulates on the feathers of these urban birds. Maybe the icy dip helps stimulate healthy blood circulation. Any number of reasonable theories might account for these birds dunking themselves repeatedly in what would seem uncomfortably cold water.

However, none of these theories captures the full essence of bathing starlings, because they exclude their experience of it. Viewing animals' lives only through an adaptive lens is like looking at bagpipes without hearing them – interesting and important, but incomplete and possibly misleading. When we imagine the experience of life for those starlings – the play of the elements against their senses, the pleasures and pains, the fears and thrills, the routines and the uncertainties of life – we regard their full splendor. When I watch starlings skipping along in their uneven trotting gait, or prying apart grass with their beaks to see what morsel might lie within, I see not only a creature beautifully adapted to a way of life. I feel the joy of a creature fully alive and crisply aware of life. Naturalist Norman J. Berrill felt it, too:

> To be a bird is to be alive more intensely than any other living creature, man included. Birds have hotter blood, brighter colors, stronger emotions ... they live in a world that is always the present, mostly full of joy.

Reaching for the animal mind

With rare exceptions, humankind has been preoccupied with viewing starlings and other animals as things, not beings. Since Aristotle first claimed that animals exist to provide humans with food and other uses, humans have held themselves apart from all other creatures. Aristotle effectively threw up a great wall between us and them. Its stones are all the ways that we believe ourselves either unique or vastly superior to our animal kin: our language, our tool use, our culture, our technology, our arts, our intelligence. Its mortar is the philosophical and religious teachings that grant us inalienable rights, exclusively give us souls, and paint us in the image of God. According to this tradition, we are his chosen species, the pinnacle of his Creation.

Cut to the 17th century, when French scientist and philosopher René Descartes debased our lowly animal cousins further by declaring that man was the only thinking, feeling being ('I think, therefore I am'), and that other animals were no more deserving of moral consideration than machines. Descartes cast out all non-human beings from the circle of moral concern with his verdict that they were thoughtless automatons incapable of thinking or feeling. These ideas fostered hideous acts of barbarity committed on fully conscious animals in scientific laboratories. With his blessing, dogs were nailed to wooden boards by their four paws and flayed alive to see the circulation of their blood. The victims' cries, to Descartes and his disciples, were no more the basis for moral concern than the creaks and groans of crushed, rusty metal.

The legacy of Aristotle and Descartes is alive and well today in the continued exclusion of animals from our inner circle of moral concern. But just as real walls begin to crumble as mosses and other small plants combine with weather to gain a foothold in the cracks, so too did cracks appear in the wall that divides man from beast. Today, the view that animals are thinkers and feelers has unprecedented scientific credibility. The change began in the 19th century, when Charles Darwin and Alfred Wallace conceived the theory of evolution by natural selection. Darwin's *On the Origin of Species*, published in 1859, offered for the first time a lucid, logical argument that placed all animals – all life indeed – on an organic continuum. Darwin went on to claim in one of his 18 books, *The Expression of the Emotions in Man and Animals* (1872), that 'the lower animals, like man, manifestly feel pleasure and pain, happiness and misery.' Unwittingly or not, Darwin removed a keystone from Aristotle's solid wall.

Naturally, this threat to the special place of humanity caused uproar and resistance, particularly from the religious community. The Scopes Monkey Trial of 1924, in which a Tennessee schoolteacher was tried (and convicted) for teaching evolution to his students, is one of the more colorful examples of this resistance. (Current debates about 'creation science' and 'intelligent design' testify to the persistence of religious opposition to evolution.) Yet so neat and unifying is Darwin's theory that it informs all branches of the life sciences, as well as other academic disciplines, including sociology, economics, and philosophy. Few scientists today – even the most religious among them – are willing to reject the idea of evolution by natural selection.

And yet, the idea that animals have minds, feelings, emotions, and consciousness is a further leap. And while many see Darwin's breakthrough as leading inexorably to that conclusion, science today is far from acknowledging it. During much of the 20th century, scientific thinking was gripped by a stifling 'behaviorist' dogma that rejected the idea of animals as conscious, feeling beings. Behaviorism rejected the study of thoughts and feelings, and asserted that the sources of behavior are the external environment, not the internal mind. Viewed as an inaccessible mystery, the animal mind was left to rot, its study deemed futile. As a consequence the recent prevailing scientific attitude towards the question of animal feelings was to either flatly deny them, or ignore them.

Accepting the case for animal pleasure requires accepting that there are sensate organisms other than humans, and probably also that they have some level of awareness. Personally, I accept that other animals have emotional feelings and conscious thoughts. As we shall see, the weight of evidence supporting this seems just too overwhelming to entertain seriously the idea that they don't. However, there are some who do not share this view. Stephen Budiansky, for instance, in his 1998 book *If a Lion Could Talk*, concludes that:

> consciousness is a wonderful gift and a wonderful curse that, all the evidence suggests, is not in the realm of the sentient experiences of other [non-human] creatures.

There remains considerable controversy in some academic disciplines (neuroscience and philosophy spring to mind) surrounding consciousness, emotions, and even sentience in vertebrate animals, and there are still scholars who staunchly refuse to accept that animals go through life with any mental experience. While it is tempting not to dignify such views with a serious discussion, it would be negligent to dismiss them altogether.

'The towering problem'

Understanding the biological basis of consciousness is a deep challenge. In July 2005, marking the 125th anniversary of its founding, the prestigious American journal *Science* ranked consciousness second in a list of 125 'big questions.' One scholar called it 'the towering problem,' and some regard it as insoluble.

How can we know whether or not an animal is conscious? We cannot. You cannot even know for certain that another human being is a conscious, thinking individual. However far-fetched the possibility, you can't know that your friends are not mechanical robots planted by extraterrestrials and programmed to respond to stimuli as if they think and feel. (I remember being caught completely off guard in *Alien* when Ian Holm's character, 'Ash,' is revealed to be an android.) Ultimately, we cannot know, and the solipsist has a logically indisputable argument. Do we accept it? Of course not. We reject it because it is so highly improbable and because it defies common sense.

Then what about other animals? To Donald Griffin, founder of the modern field of cognitive ethology (the behavioral study of animal thinking), the belief that humans are the only thinking, feeling beings on the planet is just an extension of solipsism. It is 'species solipsism,' and equally nonsensical. Says Griffin:

> nature might find it more efficient to endow life-forms with a bit of awareness rather than attempting to hardwire every animal for every conceivable eventuality.

In his 1998 book *The Evolution of Consciousness*, the psychologist Euan Macphail provides an example of species solipsism when he chooses not to accept consciousness in animals:

> it is possible to imagine computers that have been programmed to perform in a way that is just as 'clever' as some instances of animal cleverness, but still doubt whether the

> installation of cleverness alone is sufficient to guarantee consciousness.

There are at least two major problems with this sort of radical conservatism. First, as we've seen, the same could be said of other humans. Second, we are not dealing with computers; we are dealing with flesh-and-blood creatures with genes, brains and senses akin to our own, and a biological history forged in the same evolutionary workshop.

Philosopher Daniel Dennett is among a cadre of scholars who view language as critical to the emergence of a human sort of mind. Unlike Macphail, Dennett doesn't consider language as a precondition of consciousness. Macphail argues that consciousness arises only from the superb language and abstract/logical reasoning capacities of the human mind, and therefore that animals are unconscious. Dennett prefers a more common sense approach:

> The claim that, say, left-handed people are unconscious zombies that may be dismantled as if they were bicycles is preposterous. So, at the other extreme, is the claim that bacteria suffer, or that carrots mind being plucked unceremoniously from their earthy homes. Obviously, we can know to a moral certainty (which is all that matters) that some things have minds and other things don't.

For Dennett, the characteristics that make a mind conscious include: the abilities to concentrate, to be distracted, to recall earlier events, to keep track of several things at once, and the ability to notice or monitor features of its own current activities. Dennett also sees many animals as possessing what he calls 'unconscious thinking,' whereby they make and use representations but don't know they are doing so.

Neurologist Antonio Damasio from the University of Iowa conceives two broad types of consciousness: a relatively simple core consciousness, and a more nuanced extended consciousness. Core consciousness provides the organism with a sense of self in the here and now. It depends neither on reasoning nor language. Extended consciousness, by comparison, subsumes identity and personhood. An organism with extended consciousness has an elaborate sense of self, is richly aware of the lived past and of the anticipated future, and is keenly cognizant of the world beside it. Damasio sees extended consciousness as mainly the preserve of humans, though he does believe it is present at simple levels in some other animals.

Neuroscientist Jaak Panksepp subdivides consciousness into three broad levels. His primary consciousness mirrors Damasio's core consciousness – a sort of 'here-and-now' state, with raw sensory and perceptual feelings, of varying goodness, badness, and intensity or arousal. These feelings are linked to bodily events such as hunger and thirst, and to external stimuli such as taste and touch. Thus, an animal with primary consciousness would feel hunger, detect (and presumably recognize) a food-associated stimulus but with no anticipation of impending reward, and consume the food without any connection to past experience.

A higher-level, *secondary consciousness*, may reflect a capacity for thoughts about experiences, such as how external events relate to internal events. A hungry lizard who approaches a bird's nest with an unguarded clutch of eggs and anticipates the pleasant taste and the relief of hunger would be an example of this sort of consciousness. Panksepp is comfortable ascribing secondary consciousness to at least some animals. He is quite certain that animals don't think about their lives linguistically, but he believes they are quite capable of thinking in terms of perceptual images.

Like Damasio's extended consciousness, Panksepp reserves his tertiary consciousness – thoughts about thoughts, awareness of awareness, and the transformation of thoughts and memories across a linguistic-symbolic boundary – primarily for humans.

Panksepp is something of a liberal in the cautious academic field of neuroscience. He emphasizes our shared origins with other animals while acknowledging the apparently unique mental attributes of humans. He appeals to the sheer weight of 'substantial experimental evidence' favoring animals, including many vertebrates at least, as conscious, feeling individuals. Among his prime exhibits are the following:

- other mammals are attracted to the same environmental rewards, such as palatable food and social contact (as well as drugs of abuse) as humans;
- brain systems involved in human emotions share common biological origins with those of other animals;
- animals show like/approach and dislike/avoid reactions in response to artificial stimulation of deep brain regions linked to those which elicit the same reactions in humans;
- animals seem to have expectations in that they respond to barriers to success with persistence, then often frustration and anger, then giving up but rarely forgetting; and
- hundreds of self-administration experiments and conditioned place-preference studies have established that opioids are pleasurably rewarding.

Panksepp concludes that 'the denial of consciousness in animals is as improbable as the pre-scientific anthropocentric view that the sun revolves around the earth...' For Panksepp, the emergence of a brain capable of constructing ideas by intermixing input from various senses surely preceded an ability to represent things with grunts and eventually words. For example, in the conscious animal, the sound and smell of a predator (sensory inputs) should summon feelings of wariness and fear (emotions). These feelings are accompanied by mental images of a potential predator – probably a specific type, for the experienced quarry – and strategies to avoid them. This scenario also illustrates how thinking and feeling are inextricably intertwined.

Figure 2.1 A fox terrier takes off before his ball is tossed.

For me, one of the more compelling signs that an animal has extended or tertiary consciousness is behavior that suggests a transcendence of the present. In other words, if an animal acts in a manner consistent with an awareness of past or future events, then we may perceive not just a minded organism but one with extended consciousness. Anticipation is such an emotion, and as we will see later, it appears widespread in animals. There is also extensive evidence that animals plan.

They appear also to dream. The rhythmic twitchings of a dog's feet or a cat's muzzle during deep sleep are probably external manifestations of a rich imagination. Dreaming appears widespread among mammals, and recordings from sleeping birds and mammals show similar patterns of REM sleep to those of humans. When electrical activity (in the form of an EEG, or electroencephalogram) was recorded from the hippocampus (a brain area important in memory) of rats trained to solve a maze to get food rewards, patterns of nerve cell firing during sleep were very similar to those during maze navigation. In fact, so close was the correlation that the investigators could figure out where in the maze a dreaming rat was and whether or not the rat was dreaming of standing still or running.

For us, dreamscapes are vivid, and they bespeak a capacity to conjure mental events separated from us in space and time. They are the products of a mind that functions beyond mere here-and-now. And though we are obviously not awake during dream sleep, we do seem to have some consciousness of the events taking place in the mind, given our ability to remember dream fragments when we awake.

One person who has done perhaps more than anyone else to legitimize the study of animal minds and feelings is Donald Griffin. In his books *The Question of Animal Awareness*, *Animal Thinking* and *Animal Minds*, Griffin argued against behaviorist denials of impenetrable animal minds, rebuffing the prevailing view that animals (other than humans) are thoughtless stimulus-responders. He argued instead that what we see in the actions, postures, gestures and preferences of animals can tell us much about their thoughts and motivations.

Griffin died in 2003, but his influence is flourishing. Many journals are now publishing papers on cognitive ethology, including one – *Animal Cognition* – dedicated to the topic. The last decade has also seen a steady stream of new books examining animal experience and feeling. These works provide a wealth of evidence, beyond simply appealing to empathy or common sense, that planning, joy, grief, excitement, anticipation, bliss, rapture, embarrassment, esthetics, comfort, and so on, spring from minds other than human. And collectively, the evidence supports overwhelmingly the notion that pleasure is woven deeply into the fabric of animals' everyday experiences.

Even some of our spineless distant cousins are on the cognitive agenda. A recent *New Scientist* cover story carried the following provocative title: 'I Fly: Proof that Even Insects have Minds.' In another paper, two leading animal behaviorists, Marc Bekoff and Paul Sherman, urge scientists to 'investigate degrees of self-cognizance in social vertebrates and invertebrates, such as honey bees, paper wasps, [and] damp-wood termites…' More of this in Chapter 10.

When scientists, legislators, ethicists or anyone else question the existence of consciousness in other animals, it feels like the earnest quest for some elusive keystone with which to reinforce the crumbling wall dividing 'us' from 'them.' There are, of course, many aspects of existence unique to us, just as there are aspects unique to a marmot or a manatee. The human brain, coevolving with our dexterous hands, allows us to excel at things – technology, architecture, textiles, literature and lasagne, for example – that set us apart from other animals. Yet ours is still just one type of brain among many. It is not the largest (some whales' brains are much larger); nor are humans uniquely endowed with any special types of nerve cells. To date, no consciousness-producing structure or process has been found unique to humans.

It is impossible to prove beyond all doubt that animals, including humans, are conscious or even that they have emotions. But it is infinitely more credible that they can think and feel than that they cannot. There is now empirical evidence for a range of behaviors consistent with consciousness in animals from octopuses to orangutans, including concept formation, anticipation, audience effects (changing one's behavior depending on who's watching), deception, problem-solving, insight, having beliefs, and a sense of fairness. Animals are versatile in response to new challenges, they communicate requests, answer questions and express emotions. Scrub jays demonstrate episodic memory, monkeys know what another knows, and crows make tools to pry grubs from the recesses of logs.

Debates about animal consciousness and emotions will probably run and run. And as the weight of evidence continues to accumulate in animals' favor, arguments that deny consciousness to animals will hopefully become increasingly marginalized. Says veterinarian Franklin McMillan:

> Whether each feeling feels the same to animals as it does to humans – for example, does rabbit fear feel like human fear? – is not as important as the more basic issue: feelings feel pleasant or unpleasant for all species.

The struggle for existence

The cliché that animals' lives are a constant struggle for survival should have gone the way of big game hunters. Nature red in tooth and claw has come to represent the existence of all that is wild. As the narrator of a recent PBS television nature documentary about reindeer intoned, 'creatures are constantly faced with the grim prospect of "eat or be eaten," and those who migrate must endure an annual "race against time" to reach their distant breeding grounds.'

Alison Maddock epitomizes the 'do-or-die' tone in her 1971 illustrated book *Animals at Peace*: 'the ways of nature are harsh, and moments of restfulness free from struggle are rare.'

Richard Dawkins' 1995 book *River Out of Eden: A Darwinian View of Life* includes something similar:

> The total amount of suffering per year in the natural world is beyond all decent contemplation. During the minute it takes me to compose this sentence, thousands of animals are being eaten alive; others are running for their lives, whimpering with fear; others are being slowly devoured from within by rasping parasites; thousands of all kinds are dying of starvation, thirst and disease.

Little wonder we conclude that there is little peace and tranquility in the natural world. It is rather like the daily news; read enough of the doom and gloom on the front pages and you are left feeling there's a burglar eyeing your home, or that your child will be abducted on the way to school.

But this creeping paranoia is quite out of proportion with reality. For every tragic child abduction hundreds of millions of kids get to school safely every day. And so it goes for cheetahs and gazelles. Nature is not nearly so grim as she is made out to be. A gazelle, like you and I, will die only once, and that death is usually a fairly fleeting event compared to the life that goes before. A violent end on the African savannah typically lasts minutes, at most. Tens, hundreds or thousands of days precede it, few of which are punctuated by any serious threat. The same goes for gulls, sea lions, leopards, sea turtles and guenon monkeys.

Especially once an individual gets past the precarious infancy, he or she has good prospects of a long and mostly peaceful life. For every moment of fear, suffering and/or death, there are multitudes of opportunities to experience life's calmer moments and its pleasures.

'But baby sea turtles die by the score!,' I hear you cry. Sea turtle hatchlings are classic symbols of the predatory effects of rapacious nature. Of the hundred or so eggs each mother turtle lays, many are dug up and eaten long before they hatch. Of those that hatch, scores more perish during their scramble for the surf. And of those that reach the surging waters, many more fall prey to marine predators. Perhaps one in a hundred, or less, ever reaches adulthood.

I'm not challenging these statistics. I'm challenging the idea that nature is far more cruel than kind. Even an animal whose life is snatched away early still has a life. Without question there is much suffering and death in nature (our own lives are no exception). And where there is life, however short, there is the opportunity to experience it, and as long as an individual can experience penalties, so too can it reap rewards. We ought not to highlight one at the neglect of the other. When writer Robert Wright states that:

> The only thing natural selection ultimately 'wants' to keep in good shape is the information in our genes, and it will countenance any suffering on our part that serves this purpose,

he leaves out pleasure. And pleasure, like suffering, is part of natural selection.

The mass media often perpetuate the stereotype that life is harsh and joyless for wild creatures. An article on Norwegian polar bears poisoned by toxic pollutants 'migrating' from industrial regions in the south describes the 'brutal, unforgiving' surroundings. 'From the moment of birth – even conception –

animals here struggle against the odds. Most polar bears die before their first birthday.' It is sad that not all polar bears grow into adults, and a shame that humans are making things worse. And yet, a six-month-old polar bear has been suckled and nurtured by a protective mother, has experienced over 100 sunrises and sunsets, and probably hasn't bemoaned the transience of life. Most lives, even shortened ones, are probably better lived than not lived at all.

For many, interpreting nature as cruel and harsh serves an insidious purpose: the continued exploitation of animals by humans. In his 1990 book *Animal Rights Versus Nature*, ecologist Walter Howard argues that nature's 'death ethic' is so intense that anything humans do to non-human animals is better for them. Wild animals welcome death by bullet or arrow over that which nature has in store for them. He simplifies the fate of all creatures into two possible scenarios: death by human intervention or death by nature. In doing so, he neglects a far more likely fate: living to see another day.

Similarly, in a 2001 book chapter that defends laboratory experimentation on animals, American veterinarian Jerrold Tannenbaum portrays wild existence as relentlessly harsh, and undermines the possibility of happiness into the bargain:

> Most animals in the wild spend most of their waking hours engaged in the difficult tasks of obtaining food or avoiding predators. It does not seem even remotely plausible to postulate that most animals in the wild, or bred for use in research laboratories have a need or drive to be happy or to lead a generally happy life in the same way in which they have physiological needs to eat, drink or eliminate.

That predators are a constant danger does not preclude enjoyment in the would-be prey's existence. Every time we climb into a car, we knowingly place ourselves in a perilous situation. The chances of an accident on any given occasion are statistically rather remote, but cumulatively rather high. My main mode of transport to and from work for the past decade has been my bicycle. My routes have taken me along pathways and roads where motor vehicle 'predators' are a constant menace. Indeed, during my cycling career I have had three relatively minor collisions with cars, suffered many falls (one of the worst while avoiding a squirrel on a slippery pavement), and had a violent confrontation with a thorn bush (the bush won). Yet despite the risks, I delight in the thrill of getting on my bike and powering my way to and from work. The exercise is intense, and my senses are on high alert. I take in the bright colors and scents of wildflowers growing along the verges, listen for the calls of birds and frogs, and note the passage of the seasons. Even my 'predators' are a source of interest; sometimes I survey passing cars to compare the frequencies of ever-popular Toyotas versus Hondas, or to estimate ratios of sedans to sport utility vehicles (distressingly, only about 2:1 in my American suburb). And I relish those occasions when a traffic jam allows me to leave a long string of motorists languishing in my wake.

Some wildlife photographers (though fortunately a declining number) go out of their way to depict animals according to cultural expectations. I remember a 1970s television advertisement for the Ford Mercury Cougar showing a mountain lion crouched atop a high sign, snarling at the camera with mouth open and ears back in a classic feline defensive pose. Rarely do we see a photograph of a cobra *not* rearing up with his or her neck hood expanded. So overexposed is this dramatic defensive posture that when I first laid eyes on a six-foot-long Common Cobra in southern India, calmly exiting a hole in a sandbank after a rainstorm, I initially didn't recognize the species. Until fairly recently, nearly all close-up portraits of bats had them baring their needle-sharp teeth in open-mouthed horror. Merlin Tuttle, an American bat expert and photographer who founded Bat Conservation International, has helped dispel the menacing bat of popular myth by taking portraits of relaxed, non-grimacing bats.

I admit my own guilt when, having caught an adult hognose snake in Texas, I was keen to get a photo of the animal in one of its legendary defensive postures, either poised to strike in cobra-like fashion, or rolled on his back playing dead with tongue lolling out. The snake wasn't having any of it, and in any event, the most pleasing image was of the serpent woven docilely through my hands and fingers.

When animals are stereotyped, the public is done a disservice. Reinforcing the myth, we perpetuate a one-dimensional perception of the animal kingdom. Cougars are seen only to snarl, snakes to hiss, and bats to slaver. Animals experience a range of feelings and motivations, including those associated with reward and pleasure.

A subtler way that mass-media portrayals of animals manipulate our perceptions and reinforce old stereotypes is through the use of music: bassoons and basses for whales and elephants, flutes and piccolos for insects and small fish, and pounding drums during a predator's chase.

This seems benign enough, but music can be used to do more than match the size and speed of the subject. It may also reinforce notions that life in the wild is all pain and no gain. French filmmaker Jacques Perrin's 2003 documentary film *Winged Migration*, which depicts birds primarily in flight and includes footage from cameras mounted on manned and unmanned aircraft flying right in the birds' midst, is a triumph of cinematography. But the tone of the film – established by the narrative and the stirring, pulsing music – is that life is serious and earnest. One gets little sense while watching this film that birds in flight experience any pleasure or joy. Rather, *Winged Migration* is a means to escape danger – an exhausting, hazardous, life-and-death enterprise. Had the soundtrack been more light-hearted, with text that bespoke the thrill of flight and the adventure of travel, viewers may have come away with a different perception of the experience of life on the wing.

The sin of anthropomorphism

Your own feelings and motivations are a reference point for another person's feelings, and we call it empathy. Use your characteristics to interpret the behavior of an animal, and we call it anthropomorphism, which literally means *human shaped*. I once saw a remarkable photograph of a cardinal at the edge of a fishpond stuffing food into the gaping mouth of a large goldfish. An anthropomorphic interpretation of this interaction is that the bird felt sorry for the hungry fish and decided to feed her. A more scientifically palatable explanation for this bizarre occurrence might be that the bird recently lost her chicks to predation, was highly motivated to feed them, and the sudden appearance of a gaping mouth was an irresistible stimulus.

Anthropomorphism is rather *verboten* in respectable journals and conferences because scientists don't want to make declarations that can't be scientifically proven. This contributes to our failure to recognize – or at least to acknowledge – pleasure in animals. Pleasure, like sympathy, is a relatively private experience. We can't know for certain if it is present. If we interpret the behavior of playing lambs or courting cranes as pleasurable, we are being anthropomorphic because we are assuming they like it, much as we would do. And though we can't absolutely know this, it may be more reasonable to accept than to deny it. It is this appeal to reason that underlies recent attempts to remove anthropomorphism from scientists' discard pile. University of Tennessee ethologist Gordon Burghardt exhorts scientists to practice 'critical anthropomorphism,' which requires a firm knowledge of the life history, behavior and ecology of one's study species. Similarly, University of Colorado ethologist Marc Bekoff advocates 'biocentric anthropomorphism,' where we try to consider the animal's point of view and not just our own, anthropocentric view. Renowned primatologist Frans de Waal recently proposed the term 'anthropodenial' for the rejection of shared characteristics between humans and other animals.

To be fair, there are good scientific reasons for being skeptical of the uncritical ascription of human states to animals. The 'smiling face' of the bottlenose dolphin gives the impression of an incessantly happy, friendly and playful creature; it is one of the reasons for their popularity. In truth, the smile is a fairly inflexible product of dolphin skull shape, and bears practically no clue to the mood of these creatures. Similarly, the 'fear' grin of frightened chimpanzees is exploited as a happy face in advertisements and side-shows.

The terminology to be found in scientific writings provides clues as to the limits to which scientists themselves are willing to court anthropomorphism. For example, scientists would rather refer to 'rewarding' stimuli than to 'pleasurable' ones. In birds, kissing has been labeled 'beak rubbing,' and open-mouthed kissing as 'false feeding.' It's all rather sterile and businesslike. One might expect that a 1996 volume on partnerships in birds, many of which are lifelong, would disclose some examples of affectionate behavior, but 'affection' doesn't appear in the subject index, whereas there are more than 30 references to 'aggression.'

Aversions to anthropomorphism are somewhat weakened in descriptions of species whose intelligence is believed closer to our own. In his studies of chimpanzees, Frans de Waal openly affirms his use of 'anthropomorphism in its purest form,' and describes chimps as 'self-assured,' 'happy,' 'proud,' and 'calculating.' Carol Howard, an expert on dolphin behavior, applies such terms as 'disconcerted,' 'exhilaration' and 'enthusiasm' to the dolphins she observed.

If animals' lives are richly positive, or at least potentially so, then the language we use to describe them ought to reflect that. Generally, it fails to do so. In scientific parlance, animals don't 'love,' they 'bond.' The motives of those scientists who shy away from using humanized terms like 'reconciliation,' 'kissing,' and 'embracing' have been likened to 'sticking our heads in the sand to preserve our sense of dignity.' Joseph LeDoux, a prominent American researcher of animal emotion and memory, relates how he failed to get any initial academic funding for his research until he removed the word 'emotion' from his grant proposals.

Konrad Lorenz, the Austrian who shared the Nobel Prize in 1972 for his contributions to ethology, helped to usher in the modern scientific view that animals are not unconscious, unthinking automatons. Lorenz did believe that many of the behaviors he saw were instinctive, and he favored the mechanical term 'motor patterns' to describe these behaviors. Nonetheless, the terms he used to illustrate animal behavior reveal his connection with his subjects. Lorenz's intricate descriptions of social behavior in ducks include 'arousal,' 'appeasement,' 'anger,' 'triumph,' 'demonstrativeness,' 'discontentment,' 'distress,' 'incitement' and 'inquisitiveness.' Lorenz's ducks even engage in prolonged fusses, which he referred to as 'palavers.' Lorenz also recognized when the graylag geese he knew so well were feeling 'glad' and 'victorious.'

Lorenz also took care to use the term 'utter,' instead of the more mechanical 'emit' in referring to the calls made by animals. Unfortunately this failed to catch on, and to this day one still sees the latter term and rarely the former in papers describing animals' vocalizations. An animal 'that emits a sound' is René Descartes' unthinking automaton. An animal 'who utters a sound' is Donald Griffin's individual with some control over his or her communication. The latter use of language is more in concert with present day conceptions of vertebrate life. Yet the animals on the printed page continue to 'emit' their calls, and to suffer the genderless denotations of 'it' and 'that.'

That said, when I scanned a 2005 issue of the leading ethology journal *Animal Behaviour,* what I found was really quite encouraging. While some authors applied the objective term 'that' to meadow voles, turkeys, treefrogs, mangabeys (monkeys) and house sparrows, others applied 'who' to marmots and stickleback fish (the latter showed a significant preference for shoaling with friends over strangers). Furthermore, Siberian jays 'uttered more calls' in a study of nepotistic mobbing behavior.

Adherence to the traditional view of animals as unthinking objects can lead to inconsistent grammar. The caption accompanying a photograph in a large, picture book on bird behavior reads: 'A female cardinal returns to its nest with food.' Another author notes how beautifully a 'wood duck drake dazzles with its intricate color,' and an obviously male cinnamon teal preens 'itself.' So ingrained is the language that describes animals as things and not beings that even the accomplished ecologist and writer Bernd Heinrich, who routinely uses 'him,' 'her' and 'who,' makes the gaff when he notes how 'A male peacock advertises its sex and vigor by carrying around an extravagant tail...'

As animals are increasingly seen as beings and not objects, the lingo is changing, and the language of pleasure is coming along with it. Joyce Poole, author of *Coming of Age with Elephants*, believes that elephants have 'very strong emotions which could be best described by words such as joy, happiness, love, friendship, exuberance, amusement, pleasure, compassion, relief, and respect.' In *The Lives of Whales and Dolphins*, Richard Connor and Dawn Micklethwaite note the 'affection' and acts of 'kindness' displayed by these animals towards each other. Fox expert David MacDonald has written of the animals' 'playful,' 'besotted,' 'confident,' 'contented' and 'flirtatious' dispositions. Wolf watchers often include 'joyful' and 'playful' in their descriptions. Psychologist Theodore Barber, author of *The Human Nature of Birds*, refers to them as contented, happy, ecstatic, empathetic, joyous, and erotic. And Joanna Burger, a widely published biologist with over 300 academic papers and 12 books to her credit, attributes 'ecstasy,' 'pride,' 'petulance' and 'sulking' to the 45-year-old Red-lored Amazon parrot she lives with.

I'm all for scientific rigor, but when we reject even the suggestion that geese feel glad, elephants love, bats utter, fish get excited and foxes besotted, then we reduce a conscious being to little more than an organized collection of organs and tissues.

Anecdote as antidote

Much of the evidence I present for animal pleasure in this book is circumstantial or anecdotal. An anecdote is usually a single, chance observation. By comparison, data are collected in an organized fashion within strict parameters. Studies of animal behavior involve careful experiments with a statistically significant number of animals, often with both 'experimental' and 'control' groups, followed by analysis to estimate the probability that any recorded pattern was not just a chance event.

A recent study on the possible importance of tail length to mating success in male barn swallows illustrates a 'control' group. Investigators artificially created short-tailed males by trimming the tail feathers of some males, and artificially lengthened other males' tails by gluing on the clipped feather segments. The control group in this study were males caught and handled as if their tails were being doctored, but whose tails were in fact left alone. By comparing data from control birds with those of doctored birds, the investigators can cancel out any effects of handling – the stress of which can influence behavior – on mating success outcomes.

Such data may be misinterpreted, but because it is collected methodically and repeatedly it is more robust than anecdote. Anecdote combined with anthropomorphism makes for particularly tenuous interpretations. The combination carries risks of exaggeration, misinterpretation, and even fabrication.

But we should not reject anecdotes outright as useless. A telling one may be our first clue to an important phenomenon. As Bernd Heinrich puts it:

> science normally progresses one small step, one small observation, at a time [and that] discarding solid isolated observations could be tantamount to discarding critical clues.

For example, when all the members of a troop of vervet monkeys were seen to hastily flee down to the base of a tree on the approach of an eagle, ethologists Dorothy Cheney and Robert Seyfarth began to wonder how they all knew to behave in this way. These initial observations led eventually to the discovery that vervet monkeys have a small vocabulary of distinctive calls to designate different predator types. When the monkeys give the call denoting an eagle, the others hasten to the lower sections of a tree where they are out of reach. If an individual makes the 'leopard' call, the vervets disperse to outer branches that would not support a big cat's weight. The species has other calls for other threats, and it is not hard ro appreciate the importance of an unambiguous, easily interpreted set of alert signals: woe betide a monkey who thinks she heard 'leopard' and scampers to the outer branches of a tree, when it is in fact an eagle who is coming.

Many specialist scientific journals and bulletins recognize the potential importance of anecdotes by publishing 'notes,' which typically describe one or a few illuminating and provocative observations. Why ravens might fly upside down, slide down snowbanks on their backs, or swing from twigs while hanging by their beaks is open to interpretation, but these behaviors are unmistakable and intriguing.

Conclusion

It is only when we get close to animals, and examine them with open minds, that we are likely to glimpse the being within. Natural history writing is strewn with incidents in which writers are moved to awe by the intelligence, sensitivity and awareness of animals they have lived with. When Joe Hutto, a turkey hunter, lived for a year among a flock of wild turkeys in Florida, he was moved to describe them as his superiors – more alert, sensitive and aware, and vastly more conscious than himself. Hutto concluded that the birds are in love with being alive. Bryce Fraser, an Australian writer who has raised many ducks, claims he could compile a dictionary of their calls: happy cheeps, sad cheeps, panicky cheeps, desperate cheeps, when-do-we-eat cheeps, where-are-you-mother cheeps, and I'm-coming-mother cheeps.

These examples herald a paradigm shift in scientific perceptions of animals. Our past failure to even acknowledge, never mind investigate, animal feelings represents a gaping void that wants filling. Ambrose Bierce said that 'An abstainer is a weak person who yields to the temptation of denying himself a pleasure.' We have so much more to gain from an affirming view of animals – one that embraces pleasure – than from a view that denies it.

Chapter 3
FEELING SMART

The intelligence of pleasure

Is an animal less or more intelligent because it lives without clothes, central heating, and, well, atom bombs?

Jim Nollman

In our assessments of animals, humans place altogether too much emphasis on intelligence. This is convenient for two reasons. The first is that because we are highly intelligent, we automatically qualify for whatever rights or privileges we ascribe to being smart. Second, by perceiving animals as 'dumb brutes,' or at least dumber than we are, we can more comfortably distance ourselves from them morally, and continue to treat them accordingly (i.e. badly).

Is this sort of mental dominion fair? Jeffrey Masson and Susan McCarthy, authors of *When Elephants Weep*, don't think so:

> Intelligence does not imply worthiness; ... it should not matter, from an ethical perspective, how intelligent a particular species or even any other particular individual is – after all, we don't shoot a human being who is not doing as well as his contemporaries at school.

Humans have a patronizing tendency to measure the intelligence of other animals in human terms. Koko, the famous lowland gorilla and star matron of Penny Patterson's The Gorilla Foundation in California, has mastered more than 1,000 signs in American Sign Language, understands several thousand English words, and scores between 70 and 95 on human IQ tests. This score places Koko in the human slow learner – but not retarded – category. And though these results are impressive, they give a blurry picture of gorilla intelligence. They are tests devised by humans for humans, not for gorillas. And when we piously conclude that Koko actually exceeds the 'retarded human' category, the indelible image is of a beast who doesn't quite meet our lofty standards for brains.

But how would a lowland gorilla do if she were tested on things that are important to her, such as the ability to recognize native plants and to distinguish edible from inedible ones? Or to predict weather changes in the lush African jungles where lowland gorillas live? Or to gauge the moods of other gorillas based on facial expressions, body postures, or their scent? By such measures, she would be a genius among humans. Gorillas' evolutionary history and survival depend on these mental skills. Gorillas are intelligent at being gorillas. Similarly, a rat is probably no less intelligent than a rhino, but more to the point, a rat is intelligent at being a rat, as is a rhino at being a rhino.

We disparage the intellect of ostriches with our unsubstantiated stories about them burying their heads in the sand to escape detection. Do we think we could recognize the eggs laid by a particular ostrich mother among those of other mothers in their communal nests? They can, though they all appear about the same to us.

Dolphins are adept at learning to respond to computer-generated sentences, and are known to use sounds to represent 40–50 verbs and nouns. Yet by human standards, that's a pathetic word count. What wealth of information, alien to us, might be conveyed in the ultrasonic signals of dolphins? In dolphins, we have only a meager understanding of a brain that is approximately the size of our own, and that evolved to communicate in a vastly different environment.

The same can be said of the minds of other animals. Each has its own intelligence, evolved to cope with the challenges presented by their lifestyle and environment. If we want to understand the workings of an animal's mind, it helps to understand that animal's habitat, diet, mating strategy etc. – in short, its ecological 'niche.'

I don't know that fruit bats are much good at math. Their lifestyles require little or no counting ability, let alone differential calculus. But a fruit bat in a familiar forest can find a tree that's miles away and only fruits for a few weeks every other year. A Wilson's warbler could never pick its way through a tax return. Yet this bird – like so many others – is a superb navigator, migrating thousands of miles each spring to return to the same patch of woodland he or she nested in the previous year. They even recognize their neighbors from one year to the next. Dolphins were once thought inferior to primates in certain types of discrimination learning, until someone thought to test them using auditory rather than visual discrimination. And great apes were thought incapable of language until verbal training was replaced by sign language.

If we really want to understand animal cognition in its true ecological and evolutionary context – argues Sara Shettleworth from the University of Toronto – then study how nutcrackers accomplish their prodigious caching and recovery feats, not whether they can count as humans do. A single Clark's nutcracker may bury up to 35,000 whitebark pine seeds in late summer before winter sets in; and she can remember the locations of most of them come spring. Another bird, the honeyguide, can remember all of the bees' nests in a 100 square mile area, then lead a human or honey-badger to one of them.

Birds who store food by burying show not only prodigious memories that rival our own. They also show object permanence – the awareness that objects continue to exist even when they are no longer visible. Day-old chicks also demonstrate object permanence by remembering which of two opaque screens a familiar object was moved behind, then, when returned to the arena three minutes later, going behind the correct screen to find the object. This is an ability that human infants do not acquire until eight months of age. Again, this relates to life-history; having object permanence makes evolutionary sense for a mobile chick who needs to keep close to her mom or risk being lost and/or eaten. Human infants, on the other hand, are usually just beginning to crawl at eight months of age, up to which time there is no clear benefit to having object permanence.

One of the results of different intelligences is that animals may surprise – and frequently outsmart – us when we try to manipulate them to our own ends. A (pointless) smoke inhalation toxicity study exposed various rodents to cigarette smoke for four hours, five days a week. Unexpectedly, many of the animals responded by placing their feces in the smoke-delivery tubing, repeatedly and in quantity. One poor hamster stuffed the air inlet so effectively that he/she suffocated.

No bird is more maligned by humans than the domesticated chicken, *Gallus gallus*. It is not enough for us to subject tens of billions of chickens each year to factory farms and slaughterhouses, we deepen the injustice by denigrating them at every turn. A coward is said to be 'chicken.' Chickens and turkeys have also become emblematic for stupidity – witness the myth that turkeys drown in torrential rains by standing open-beaked with their heads turned skywards (they don't).

A closer look at these birds reveals that the stupidity lies not with them but us. Despite their brainless reputation, chickens have 25 to 30 different call types (at least that's how many humans can distinguish – the birds probably recognize many more). Chickens also have the brains for deception. Roosters curry favor and increase mating opportunities with hens by announcing the presence of a discovered bit of food, such as a seed or a juicy caterpillar. I have seen roosters do this on a farm near my home; to watch the hen come running, then watch her eat the morsel that he might have enjoyed for himself is to witness chicken chivalry at its finest. But roosters are not always so gallant. They sometimes deceive hens by uttering the food call when there is no nice tidbit. It is thought that this behavior may occasionally be rewarded with a mating. Obviously though, it would not pay to deceive too often, for soon the deceiving rooster may be recognized as a faker and shunned by hens. Studies find that roosters only utter deceptive food calls when a hen is some distance away and easily fooled by the ruse.

I think that it stretches credibility to think that such sophisticated behaviors could be instinctive.

Yet that is what some claim. Behavioral neuroscientist Lesley Rogers provides a good example of scientists bending over backwards to avoid interpreting an animal's behavior as intelligent. In her 1997 book *Minds of Their Own*, Rogers describes a study published in 1981 by B. F. Skinner (champion of the behaviorism we encountered in Chapter 2) and colleagues at Harvard University, who trained pigeons to locate colored spots placed on their bodies by pecking at them. Once the pigeons had learned this, they were fitted with a bib, beneath which a dot was placed so that it could only be seen in a mirror. When the birds noticed the reflection of the dot, they did not peck at the mirror, but uncovered the dot on their breast and pecked at it. This finding strongly suggests to me that pigeons are self-aware, an aspect of consciousness that had been demonstrated, to much fanfare, in chimpanzees, and which spawned much debate and controversy that continues today. Rather than interpreting the pigeons' abilities as a sign of higher than expected intelligence, the authors concluded that 'if a bird can do it, it cannot be complex behavior and it cannot indicate self-awareness of any sort.'

Twenty-five years on, such conclusions are thinner on the ground. Distinguished ethologist Peter Marler from the University of California at Davis recently reviewed social cognition in non-human primates and birds and concluded that there are more similarities than differences between birds and primates. Size is not everything, and this applies also to brains. Despite their relatively small brains, Rogers also concludes that birds can perform at the cognitive level of primates. Bird brains differ from mammal brains in having no neocortex (the highly developed outer portion of the mammalian brain). Their intelligence arises from the proliferation of tissue in an entirely different part of the brain, the paleocortex. In early 2005, an international consortium of avian brain experts announced a total makeover of the naming of bird brain parts, replacing terms that implied primitive features with terms that reflect a recognition that birds process information in much the same way as does the vaunted human cerebral cortex.

Smart sharks and flexible frogs

When we recognize that animals evolve different sorts of intelligence for tasks that are important to them, we may expect them to exceed us in some forms of intelligence. This shouldn't humble us; it's just nature. Dogs smell better than us (unless they have been rolling in dead fish), so they have higher 'olfactory intelligence.' Bats detect and interpret echoes with a precision that our best sonar equipment can only hope to emulate. Compared to them, we are ultrasonically challenged.

Pigeons are better than humans at recognizing objects that have been rotated at different angles. Perhaps this reflects a bird's-eye view of the world, where familiar landmarks are regularly spotted from different angles depending on flight path. A group at Oxford University recently reported that some pigeons use their knowledge of human transport routes to navigate. They turn off at certain motorway junctions and use landmarks to remember where they are. Study leader Tim Guilford said: '... when they do follow a road, it's so obvious. We followed some which [sic] flew up the Oxford bypass and even turned off at particular junctions. It's very human-like.'

A week before I read about this study, my wife and I were sitting in a coffee shop in downtown Liverpool. The window in front of us faced a long city street, and we couldn't help noticing when a pigeon flew towards us right down the center of the street, then made an abrupt right turn at the crossroads. One possible benefit to following roads is that it would free the pigeon's brain from consciously doing a difficult task (navigating using subtle cues), so allowing them to focus on other tasks, like watching out for predators.

Guilford's team believes that the pigeons use their more sophisticated magnetic and solar compasses to navigate when they are over unfamiliar territory, but that when they know the area well, they follow familiar landmarks, making diversions to follow roads home. Thus, this study not only illustrates pigeons' intelligence and resourcefulness, but also their awareness and flexibility.

The great white shark offers another fine example of the chasm between popular perception and reality. The perception, flogged endlessly by books and films, is that these huge creatures are killing machines, indiscriminately eating practically anything they encounter. As a wide-eyed teenager, I remember reading an improbable list of items (including a rubber tire) found in the stomachs of great white sharks who had been eviscerated by humans.

Recent studies of great whites show these inedible fish to be less savage than the fishermen who haul them onto their boat decks and pose for the camera as conquering heroes. As one might expect of a large vertebrate that grows to over 25 feet and lives for decades – the third-largest of all fish species is now understood to be intelligent, and to have individual personalities.

The brain of an adult great white shark is narrow, but nearly half a meter long, with a highly developed cerebrum, a brain area associated with learning, memory and thought. In some vision tasks, sharks have been found to learn many times faster than cats. By and large, humans are undesirable and unpalatable to sharks, and on those extremely rare occasions when a human is attacked by one, it is believed to be out of curiosity or mistaken identity (a surfer laid out on a surfboard can look uncannily like a seal from below). And it is because human skin is so relatively thin that shark bites tend to be so serious. Nevertheless, only 30% of great white shark attacks prove fatal to humans. (About 50 to 75 people are attacked by all sharks worldwide each year, of whom about 5 to 15 die.)

Some shark species are believed to hunt and feed cooperatively (or at least coordinatedly), and to eavesdrop on dolphin echolocation to locate good food sources. And like fruit bats and orangutans who know when and where individual trees come into fruit in their extensive jungle habitats, great white sharks appear to maintain internal timetables of seasonal banquets located thousands of miles apart. They reach these with an equally sophisticated navigation system using, perhaps, memory maps, aquatic corridors of scent, or their electroreception of the

Earth's magnetic field. Recent study suggests that spatial behavior in fishes is as complex and elaborate as that of land vertebrates, and that it is based on learning and memory processes arising from brain circuits that may be homologous to those of mammals and birds.

Sharks also use body language to communicate moods and motives with each other. At least nine distinct body postures have been detected so far in one species of hammerhead shark.

Plainly, animals are not pre-programmed for all the things they do. To live in a complex, changing world it pays to be flexible. Machines are not flexible. They have the advantage of performing what they were designed to perform very efficiently. But change the rules and they are lost. Replace the gasoline in your car's tank with any other liquid and you won't get far. Feed a piece of thick cardboard into a laser printer, and while you may have a minor disaster on your hands, you certainly won't have a print-out. Put a green juniper berry among the frozen peas fed to my three pet rats, and you'll later find three sated rats and only the berry left in the cup. When a starling flies in a straight line over a field, are we to believe she has no idea where she is going? Or should we assume she has a specific goal in mind, such as arriving at her nest, a favorite perch, or a ripe fruiting bush? When a vole is spotted on a bird-feeder grabbing larger tidbits and flinging them over the edge, then quickly climbing down and retrieving them from the ground below, should we believe these are only the robotic movements of a creature with no foresight? And when an enterprising sheep solves an eight-foot hoof-proof cattle grid by rolling over it, should we assume it is just a happy accident? If so, then how to explain that the rest of the flock soon learned the trick, and regularly rolled themselves across the grids to raid villagers' nearby gardens?

Donald Griffin saw the adaptability of having a mind:

> Consciousness confers enormous advantage by allowing animals to choose to act in ways that get them what they want or

> avoid what they fear. Thinking animals can try out possible actions in their heads without risk of actually performing them solely on a trial-and-error basis. Considering and then rejecting a possible action because one decides it is less promising than some alternative is far less risky than trying it out in the real world, where a mistake can easily be fatal.

Planning and thinking are immensely adaptive. Animals who make bad decisions pass fewer or no genes into the next generation.

Even reptiles and amphibians, so often dismissed as inflexible automatons, show signs of awareness. Hognose snakes, native to North America, take longer to emerge from their 'death feign' display in the presence of a (stuffed) owl or an attentive human than with no owl or a human whose eyes are averted. Some lizards raise and wave their tail after spotting a nearby predator. As Gordon Burghardt interprets it, the message in effect is: 'I see you. Don't waste your time and energy trying to catch me for I am fast and you've lost the element of surprise.'

If the small ponds in which African bullfrogs spawn begin to dry up, the tadpoles begin calling. The attentive male frog responds by digging a trench through the mud from the nearest pond to the imperiled tadpoles, thereby creating a channel through which the stranded tadpoles can swim to safety. As no two effective channels could be exactly the same length and direction, this necessarily flexible behavior seems also to be guided by some degree of planning and cognition. In his writings on animal minds, Griffin even challenged the assumption that an insect's brain is too small and its central nervous system too differently organized from ours to be capable of conscious thinking and planning or subjective feelings.

The forgotten individual

If we are to appreciate animals as having pleasurable lives, it helps also to view them as individuals. Too often, animals are seen merely as members of a species. In many animals, inter-individual differences are comparable to those of humans, though to us they may all look the same on the surface. Ravens clearly can identify each other as individuals, yet even the most expert human cannot. This says something not only about perceptual acuity on the part of the birds, but also about the greater amounts and range of relevant sensory information available to those of their own kind than to us.

Sheep can recognize 50 or more members of their flock from photographs of their faces alone, and retain this ability for two years. They can even recognize familiar faces when presented for the first time in profile. By overlooking the physical and mental character of individuals, viewing all sheep merely as members of the species *Ovis aries* is rather like viewing all of Cezanne's paintings as if they were the same.

Superficially, all gray squirrels, starlings, and iguanas may look the same from our usually limited vantage point, but they are not carbon copies. Whenever biologists study them intensively for long periods, they invariably come to recognize each as a character with distinct personality traits. As Peter Steinhart puts it in *In the Company of Wolves*, 'There are soft wolves and hard wolves, kind wolves and malicious wolves, soldiers and nurses, philosophers and bullies.' Octopuses have been described in the scientific literature as having different personalities: passive, aggressive, and paranoid.

There is even anecdotal evidence that insects have something akin to individual personalities. Nature photographer Hugo Van Lawick, who settled in Africa with Jane Goodall, wrote of taming mantises and jumping spiders to sit on his hand and take food from his fingers, and described the eight insect pets he had at one point as a distinct individual with its own character.

A recent study found that spatial foraging behavior in a colony of desert ants is idiosyncratic to each ant. This might not quite amount to personality, but it reminds us that variation is what evolution by natural selection acts on for all organisms.

Ironically, owing to their unconventional reproductive systems, worker ants of a colony share more of their gene complement with each other than do siblings of most other species. Perhaps studies of less closely related conspecifics will yield greater individual variation than among the social insects.

Pleasures, too, are individual. Those who rehabilitate sick or injured animals before releasing them back to the wild soon come to know each as a unique individual with distinctive preferences. One of nature writer Mike Tomkies's tawny owls liked to bathe in his dishwashing tub. Another chose for her perch a stag's antlers that he had found and fixed to the wall. Of the three rats I adopted from a local RSPCA animal shelter, Veronica is much more likely to be on the deluxe solid-floor exercise wheel than is Rachael, who prefers to play games of hide-and-seek under an old curtain remnant strewn across the floor. Lucy insists on using the solid-floor running wheel as a toilet, with the result that exercise bouts can become quite noisy towards day's end, as dried pellets go for a spin.

I recently saw a goose do something I had never – in 30 years of birdwatching – seen a goose do. A graylag repeatedly dived and swam two feet under the surface, as if doing her best imitation of a cormorant. Curiously, a male mallard (a dabbling duck also not known for diving) was doing the same thing a few feet away. It looked like two birds having an identity crisis. They evidently were not trying to escape some perceived danger overhead, for there were other geese resting nonchalantly on the riverbank just 20 feet away.

Because we live so closely with them, we are more likely to witness such rare moments in our pets – or, to be more politically correct – companion animals. Most who live with cats or dogs will have experienced occasions when the animal manifests some unusual behavior that we might never have known they had in them. One morning, for example, I noticed our two-year-old neutered male cat Camille stalking away from our front storm window door looking agitated. The fur down his back was erect and his normally slender tail looked like a chimney-sweep's brush, as if he had just seen another cat or a dog or something else that upset him. Two minutes later, my eight-year-old daughter yelled out 'Aieee! Camille just peed on me!' Sure enough, the cat had sprayed against Emily's back while she was tying her shoelaces. This had never happened before, and it never happened again. As Emily changed her clothes upstairs, I mollified her with my tentative theory that Camille had seen something threatening and that he had reacted by marking something on his territory that he regards as precious and rightfully his.

Incidents like these remind us that animals are not predictable, one-dimensional creatures. They are sensitive and complex. Like us, they have temperaments and moods, and their individual personalities express themselves in diverse ways. A goose does not always behave goose-like, a sheep has her special friends, and some octopuses blush easier than others.

The star-nosed mole's nose knows

Physical pleasures are sensory. When we enjoy a feather against our skin, our touch receptors allow us to experience the sensation. When we breathe in the sweet scent of an apple or the rich aroma of freshly ground coffee, our sense of smell delivers the reward. When we admire the symmetry of a butterfly's wings, or a blush of poppies on a hillside, we are responding to signals from the rods and cones in our eyes.

Some forms of sensory intelligence in the animal kingdom involve extensions of senses we experience, such as the ability to see ultraviolet light, to hear at higher or lower frequencies, or sensitivity to smells or tastes at chemical concentrations we cannot detect. Human hearing spans 50 to 20,000 Hz. Mice hear sounds mainly between 10,000 and 100,000 Hz. Dogs hear up to 45,000 Hz, cats to 85,000 Hz, dolphins to over 100,000 Hz, and bats to over 200,000 Hz. Elephants hear as low as 8 Hz. Crocodiles, once thought to be silent, are now known to communicate using infrasound vibrations transferred through the water. These massive reptiles have pressure sensors all over their bodies that detect these signals. The sensory surface of a typical dog's nose is some 25 times that of a human's nose, and the dog's brain has many thousands more olfactory cells (7,000 mm^{-2} compared to 500 mm^{-2}). A polar bear's sense of smell is believed to be some seven times more sensitive than that of a bloodhound. The oceanic white-tipped shark may detect blood at concentrations as low as one drop in an Olympic sized swimming pool. And there is some evidence that dolphins might detect that their human teachers are pregnant before the women themselves know it.

Many animals possess senses with which they can perceive their environment more acutely than can humans. Being able to perceive certain airborne fruit or blossom smells is more important to fruit bats than to humans. Moles are acutely sensitive to vibrations in the soil. Spotting a flying midge from twenty feet is something that fly-catching birds and birds of prey are better at than humans. Because they involve the brain, sensory abilities are a form of intelligence. We might loosely call it 'sensory intelligence.'

Other forms of sensory intelligence fall outside our experience, such as the capacity to navigate using geomagnetic, celestial and chemical cues, or to communicate with and detect electric charges. Golden moles from Africa sense termites by the vibrations they make in the sand. Indeed, many animals have seismic sensitivity, ranging from spiders, scorpions and insects to frogs, kangaroo rats, elephant seals and elephants. The star-nosed mole's nose is so well served with tiny nerves that 600 could fit on the head of a pin. Pintail knifefish of South American rivers have 80 electric receptors per square millimeter (an area equivalent to the tip of a ballpoint pen) around their head. Until recently, the way a seal tracks and catches fishes in the total darkness half a kilometer below the ocean surface was a mystery. We now know the seals detect the subtle turbulence left in the wake of a swimming fish, which takes 3–5 minutes to dissipate.

So, might animals with more acute senses than ours be capable of greater pleasures?

At least three contemporary moral philosophers have entertained the idea that animals could be capable of more intense feelings (including pleasures) than us. Imagine suddenly being endowed with ten times the number of nerve cells you have for smelling things. How then might a favorite food or a flower's perfume smell? Imagine being able to see the ultraviolet light spectrum, as many other animals can do. A field of spring flowers, a school of fish, or a flock of birds might pulsate with a different sort of intensity from what we experience. The plumages of some birds are more vivid and bright in ultraviolet light, which other birds, unlike humans, can see. Ultraviolet light reflectance is found in the plumage of most birds.

When I hear a wren belting out his song from a nearby shrub, I wonder: how might it sound if it were stretched out to 30 seconds, allowing me to detect and appreciate all 100 notes compressed into his 8 second aria? Or the cowbird, who packs 40 notes into his one-second serenade? As a graduate student studying vocal communication between mother Mexican free-tailed bats and their pups in southern Texas, I would record at high speed and analyze the calls by playing back my recordings 30 times slower. The baby bats' repeated calls to their mothers, each too high for me to hear and lasting barely a tenth of a second, transformed themselves into repeated cries. Each baby's call was temporally stable and distinctive, ensuring that the mother could recognize them among the millions who roost in their vast creches during the month-long nursing period.

A heightened capacity to detect sensory stimuli does not necessarily mean that an organism *experiences* these stimuli with greater intensity. Nor do sensory abilities necessarily have relevance to experiencing pleasure. But some – those relating to food, sex and esthetics, for example – probably do. That an organism detects and responds to certain sensory cues indicates that those cues are biologically meaningful to that organism.

Nor can we assume that what fails to please us fails to please them. It would seem a cruel twist of evolution if turkey vultures actually hated the smell of carrion (though according to humorist Frank McKinney Hubbard they do abhor discovering a glass eye). The same goes for dogs and jackals, who take much pleasure in rolling and rubbing their necks and backs on carrion.

Should we assume that another species can, at best, only feel pleasure as intensely as we humans can? Joseph Wood Krutch was wary of this assumption when he wrote:

> It is difficult to see how one can deny that the dog, apparently beside himself at the prospect of a walk with his master, is experiencing a joy the intensity of which it is beyond our power to imagine much less to share.

Furthermore, let's not assume that humans experience the full gamut of emotions to be found in the animal kingdom. In light of their sometimes vastly different living circumstances and sensory capabilities, other species may experience some emotional states that we do not. Joyce Poole, who has studied African elephants in Kenya's Amboseli National Park for more than 25 years, is confident that while humans and elephants experience many emotions in common, elephants experience others that we may never know.

Onward pleasure

The fuss about animal intelligence is a bit of a canard. First of all, it has only limited relevance to matters of feeling. As 18th-century philosopher Jeremy Bentham famously said of animals: 'The question is not: Can they talk?, or Can they reason?, but Can they suffer?' We may concede that some animals – and I'm thinking here of very small ones, like amoebas, or relatively simple, sessile ones, like sponges – are not sentient. We should not assume that more complex, mobile ones with nervous tissue, like land-snails, earthworms, and damselflies, are devoid of feeling. And while it may be difficult to assign sentience to invertebrates, it should not be so with our more familiar cousins in the pantheon of life, the vertebrates. These creatures all have nervous systems, a suite of senses by which to perceive their worlds, and they all move about. Evolution has endowed each with the tools for avoiding pains and perils, and for securing rewards.

We must put aside the prejudices we hold towards other creatures. They are built on two thousand years of pious presumption that humans are the chosen ones, inexorably walled-off from all the others. Nourished by the inescapable knowledge that we are evolutionarily continuous with the other beasts, we are now realizing – from scientific study and empathic observation – that ours is a planet rich with other minds and experiences.

We can never actually share the mental experience of another creature, including another human. But we *can* and do ask questions, and design clever experiments in which animals can demonstrate their perceptual thresholds, memories, sensory preferences, individuality and their personality traits. In the face of these discoveries, the position that pleasurable states are the sole domain of the human species is narrow and anthropocentric. To deny animals conscious experiences is to deny that they plan, desire, anticipate, tease, grieve, enjoy, tolerate, and gauge. It is to reject that they make decisions.

Part I has addressed why we should expect the widespread existence of pleasurable experience in the animal kingdom. Its purpose has been to provide a foundation – a basis for predicting that animals, like us, feel good. However, predictions are not proof. Part II presents the direct evidence. Ultimately, there can be no decisive proof of animal pleasure, any more than there can be absolute proof that smoking causes lung cancer, or that bacon is bad for you (it's certainly bad for pigs). But just as the accumulation of evidence indicates, convincingly, that cigarettes are a leading cause of cancer (and statistics), so too does the accumulation of evidence persuasively show that animals feel good.

Whether or not you are satisfied that animals are thinkers and feelers, many people continue to balk at the idea, and it still behooves us to look at the evidence. What can we discern from the way animals behave? How do they respond to stimuli that we humans typically find pleasurable? When animals play, are they having fun? Do animals eat merely to survive, or do they savor their food? Is sex pleasurable for animals? Do they like physical contact? Let's see.

Part II
WHAT ANIMAL PLEASURE?

Chapter 4
PLAY

Fun for its own sake

Man has always assumed that he is more intelligent than dolphins because he has achieved so much – the wheel, New York, wars and so on – while all the dolphins had ever done was muck about in the water having a good time. But, conversely, the dolphins had always believed that they were far more intelligent than man – for precisely the same reasons.

Douglas Adams, *So Long, and Thanks for the Fish*

Who hasn't seen the way a dog will drop a ball at the foot of a human companion and wait, head cocked, in anticipation of the next toss? Can anyone doubt that the animal is playing, and enjoying it?

Of all the behaviors animals engage in that we might interpret as pleasurable, play is the least scientifically controversial. A fairly straightforward academic definition of animal play goes thus: 'any purposeless motor activity that includes patterns from other contexts such as mating or stalking.' Open any textbook on animal behavior, and you will find references to play. There are numerous scientific articles and books dedicated to animal play. It is usually easy to recognize and it is not reserved only to animals who eat food from a can. Virtually all young mammals as well as some birds play, as do the adults of many species. Nevertheless, play is poorly studied in animals. In 1998, animal play expert Marc Bekoff lamented:

> Incredibly, we still do not know if most mammals do or do not play, and our knowledge of the distribution of play in other vertebrate classes [birds, reptiles, amphibians and fish] is weaker still.

Though play is undeniably adaptive, it is pleasure, curiosity, and joy that provide the motivation for play in animals and humans alike. Play is a good indicator of well-being. It occurs when other needs, such as food, shelter and safety, are sufficiently met, and when unpleasant feelings like fear, anxiety and pain are minimal or absent. Otherwise the animal's efforts would be directed at meeting these needs or relieving these feelings, at the expense of play.

Play serves many functions that may help an animal to survive and succeed in life. These include: developing physical strength, gaining skills, acquiring knowledge, learning the ropes of social behavior, and exploring.

This is probably why it evolved. But when a pair of mountain goat kids chase each other, jumping, twisting and kicking, they are hardly training for becoming good grown-ups. Animals play for fun, not for keeps.

There are at least four good reasons to believe that animals are having fun when they play. First, they look like they're having fun. Cats chasing pulled string, young squirrels romping, or otters sliding down snowbanks look like they are heartily enjoying themselves. I remember watching three little eastern gray squirrels romping and wrestling around the base of a palm tree in Orlando, Florida. They leapt on and off the bole in pursuit of each other, and at times fused into a single squirrel ball as if they were one.

Second, humans enjoy playing, and much of our play resembles that of other animals. There is an element of *funktionslust* in the playing of sports. We usually put our all into it and strive to do our best. We tend to favor games we are good at, and performing well is one of the rewards. Animals may get similar pleasure from their play because it invariably involves doing and refining things they are good at. The play of young predators commonly involves chasing and catching things, such as an adult's tail or flying insects, and wrestling with each other. The play of herbivores, such as young goats, entails leaping, running and zigzagging, skills useful for negotiating difficult terrain and evading predators.

Third, animals want to play. In the laboratory, young chimps and other species will play rather than eat unless they are very hungry. Some animals will work for the chance to play. Junior, a captive orangutan at the Saint Louis Zoo, would clean up his cage in return for the opportunity to play with his whistle. The urge to play can be irresistible.

Fourth, there are chemical changes in the brains of playing animals that suggest they enjoy it. Rats show an increase in dopamine production in their brains when anticipating opportunities to play. Jaak Panksepp reports a close association between opiates and play, and that rats enjoy being playfully tickled (see Chapter 7).

Because play often mirrors serious, dangerous interactions such as fighting, attacking prey or escaping predators, it is important that individuals recognize the playful intentions of others. Most animals have body language to signal their desire to play. Dogs use a 'play-bow,' in which the soliciting animal faces her playmate with forelegs flat on the ground and hindquarters raised up, tail usually wagging. The facial expression is relaxed and 'smiling.' Tail wagging in dogs is without doubt a means of communication. Dogs wag their tails when receiving food in the presence of humans, but not when they are videotaped by a hidden camera.

Other species use play signals too. In chimps, baboons, and other primates, the invitation to play is accompanied by the 'play-face,' in which the mouth is opened wide but the lips hide the teeth. It conveys the important message: 'this is not serious, even if it may seem so.' Bernd Heinrich describes the raven's 'play walk': a bird's back is rounded, its head slightly bent down and pulled back between the shoulders, and its movements exaggerated. Arabian babblers signal their desire to play by crouching, rolling over, picking up sticks, with play bows, or by

Figure 4.1 Ethiopian wolf pups play at fighting.

establishing eye contact. Lambs invite other lambs to play by leaping in the air and kicking out their hind legs. Coyotes do a strange, squirming roll, mongooses whip their tails, foxes make little running movements with their hindlegs, badgers use head-twists, polecats a stiff-legged jump, and brown bears a kind of head-wobbling.

Animals sometimes play alone, but more usually it is a social activity. As such, fairness is also important when animals play, as when humans do. Play involves cooperation and give and take, and requires that players agree to participate. That play rarely escalates into all-out fighting is because animals abide by the rules and expect others to do likewise. Video studies at the University of Lethbridge, Canada, found that playing rats monitor one another and then fine-tune their own behavior to maintain the playful mood. Red-necked wallabies calibrate the boisterousness of their play to suit the age of their playmate. Marc Bekoff believes that animals play in part to develop a sense of morality. When individuals roughhouse, they form social bonds and learn what is acceptable: how hard they may bite, how roughly they can interact, and how to resolve conflicts.

Play is most common in young, but many adults also do it. Male cheetahs often form lifelong coalitions, and the members of these small groups of two or three will play together throughout adulthood. In a laboratory study, juvenile and adult male rats preferred a box containing a 'free' moving rat compared to

Figure 4.2 Young rats at play.

either a box with a rat confined behind Plexiglas or a box with no rat. The confined rat was visible, but not available to play with.

Rats are not the only rodents to engage in play. So too do black-tailed prairie dogs, Columbian ground squirrels, alpine marmots, olympic and hoary marmots, house mice, gerbils, Chinese and golden hamsters, and chipmunks and other squirrels. Play-chasing and escape play occur in several species, including the African giant rat, African ground squirrel, and green acouchi. Other rodent play behaviors include crawling under/over siblings, mouth-wrestling, and sexual play. House mice, particularly juveniles, engage in both locomotor and object play. There is also evidence that greater environmental complexity and space encourage laboratory-housed mice to play. Adult male house mice put in groups of four in spacious glass terraria with ladders, platforms, a climbing tree and ropes leap and hop much more than mice housed in commercial laboratory cages (just 38 × 22 × 5 cm) with fewer toys.

Roscoe, one of the cats I have had the pleasure of being owned by, is a fine study of the value of play to an adult cat. At two years of age, Roscoe had the misfortune of losing two elderly feline companions in fairly short succession, one of whom he was close friends with and played with regularly. A family situation necessitated his being a solo cat for many months. I had often played with him since he was a kitten, and with his loss of feline playmates, I became the sole target of his demands for fun. Roscoe is without question a toy-snob, rejecting those plastic toys and jingling rubber balls whose purpose seems more to accumulate under sofas than to amuse cats. Roscoe's commercial toy of choice is the Cat Dancer, a length of high-tension wire with little bits of tightly wrapped brown paper on the end. He will also chase the end of a long piece of string (preferably dental floss), and in bursts of intense friskiness, will pursue an unshelled hazelnut rolled across the floor, or leap high into the air if I toss it.

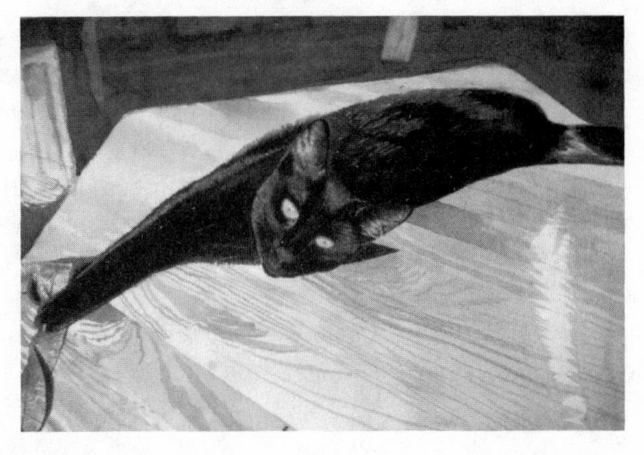

Figure 4.3 Author's painting of Roscoe toying with the author's tie.

Roscoe uses a number of devices to get me to play. He will meow insistently while fixing me with a stare, or make staccato barking noises at the ceiling. Another ploy is to prance into the room sideways, back and tail arched, stotting on all fours as a kitten may do in confrontation with a larger foe. Obviously, Roscoe is not afraid of me and has no need to try to scare me by looking big. He is simply getting my attention. These efforts often go rewarded, and there is probably an element of conditioning going on here. But conscious intent too is eminently possible. If Roscoe's thinking could be expressed in human terms, I think it might go something like this: 'I want to play. Jonathan often plays with me. Let's get Jonathan's attention and try to get him to pull that Cat Dancer for me. "Meow!"' Playing has little survival value for a cat with food and shelter. He plays for the pleasure in it, as we humans do when we play tennis, or Ultimate Frisbee.

Gravity games

Gravity is such an inexorable force that ways to defy it are exhilarating. In Iqualit, an arctic village north of Hudson Bay, Canada, locals have spied ravens hanging upside down, swinging up onto and somersaulting over powerlines, hanging from them with their bills, and sliding down roofs. A woman there described a group of ravens taking turns rolling down her roof, either flying or walking back up to roll down again. On another occasion, two ravens played a form of 'rodeo' on two loosely strung, wind-whipped overhead power lines. They took turns trying to grasp the second wire in the bill and hang on as long as possible.

There are numerous similar accounts of raven antics in scientific journals and books. A notable trick is flying upside down, sometimes for 100 meters or more. Raven authority Bernd Heinrich has watched Houdi, one of the birds he raised from a chick, sliding and rolling on her back repeatedly down a two foot high snow mound. Ravens have also been seen sliding on their backs down snowy slopes in both the USA and Britain, and on their breasts in Maine.

Hiking near Kenmare, Ireland, in the late 1990s, Sheila Key of the Department of Botany and Plant Sciences at University of California at Riverside noticed a group of jackdaws (another member of the corvid family to which ravens belong), perched on a telephone wire. Key shared with me her observation:

> He (she?) would hold on to the wire and let himself fall backwards, letting his body hang down and relaxing his wings so they hung down from the body. He would look backwards at his companions on the ground. After about 5 seconds, he would let go and fall off the wire, then fly back up to sit on the wire and do it all over again. He did this three times while we were watching. It had to be nothing but pure play.

In 2003 a BBC documentary on Australian wildlife featured footage of one or more corellas – a gregarious white parrot – hanging from twigs by the beak, and upside down by the feet.

Sliding down things is a popular activity for many animals. Snowbanks and muddy or grassy slopes provide tobogganing opportunities for penguins, otters, ravens, and bears. Alaskan Buffalo have been seen sliding on ice. Even young captive alligators repeatedly slide down slopes into water.

Man-made structures and objects afford many opportunities for gravity games. Keas – large New Zealand parrots – enjoy sliding down the slope of a car's rear window, and crows slide down the Kremlin roof. In Japan, carrion crows have been spotted sliding on solar panels, their wings partly spread for balance, and down a children's slide standing and on their sides with wings tucked in.

Tiko (Joanna Burger's 45-year-old Amazon parrot) took to sliding down the stairway banister, breaking perilously near the bottom by tightening his grip. He even took pains, and apparently pride, to show Burger his new skill, luring her from the kitchen with cooing sounds, climbing to the top, and skiing down as she watched in delight.

Ellen Marsden, biologist and lecturer at the University of Vermont, described to me the sliding game of a tame river otter she once watched while attending a wildlife conference. One of the attendees would exercise the otter in the broad, carpeted hallways of the hotel. He would walk about 20 feet away from the otter, who waited behind. On signal, the otter would bound towards the man, then at the point of passing him, tucked all four legs back and performed a long belly slide across the slick carpet, coming to rest some 20 feet further down the corridor. It's not a gravity game, but it probably has its roots there.

Many birds exploit gravity to get at food. Egyptian vultures drop stones on ostrich eggs to crack them open and get at the contents. Lammergaiers (large vultures of European mountains) drop large bones onto rocks to crack them open to access the nutritious marrow inside. Gulls use similar tactics to crack open shellfish, and crows with nuts. It might not be a big evolutionary step from dropping objects to get food to dropping objects for fun and/or mischief. Groups of 15–20 ravens play a form of aerial tag with a stick; when one drops the stick, another catches it and becomes 'it.'

Members of the swallow family often toy with feathers. The BBC television documentary *The Life of Birds* includes brief footage of purple martins doing just this. One bird carries the feather up, drops it, and others attempt to catch it. Birdwatchers have spotted the same sort of thing in kestrels. Patrick Kramer of the Purple Martin Conservation Association, in Pennsylvania, points out that these aerial insectivores may instinctively pursue objects dropped by other martins in hopes of a meal or nesting material. But it may also be fun for the birds.

Evolutionary biologist Sylvia Hope, with the Department of Ornithology & Mammalogy, California Academy of Sciences, recalls watching apparent play in a mixed flock of western and glaucous-winged gulls on the bay-shore in Berkeley, California. One bird would fly high in the air and drop a stone; others would swoop to catch it. This continued for some minutes. It resembles opportunistic foraging behavior sometimes performed by gulls while they are fishing at sea. A flock gathers above a school of fish, birds dive in to retrieve a fish, and the winner quickly rises again to make off with the prize. Other gulls may try to rob the prize-holder, and in the ensuing fray, the fish is often dropped and caught in air by a successful pirate.

Gulls are certainly not all business. In autumn 2003, I watched black-headed gulls on the River Ouse, in Yorkshire, plucking leaves from the water surface, holding them in their beaks, and flying with them. One bird might drop a leaf and fly down after it, calling. Sometimes two birds chased each other on or just above the water. Other birds dipped and dived in flight to scoop leaves from the surface. This frolicking had no obvious or immediate survival purpose. A three-year study of similar aerial drop-catching behavior by Herring Gulls in Virginia concluded that it was play, based on several factors: drop-catches were performed more by younger birds, were not necessarily made over hard ground (a tactic the birds use to crack open clam-shells), and were sometimes performed with non-food objects, and more often during warm, windy weather.

Bats can be highly responsive to objects tossed into the air. As a member of a team studying the foraging behavior of red bats in southern Ontario in 1985, we exploited this to lure bats into mist-nets by tossing small pebbles strategically into the air as they approached. In the case of small pebbles, one might readily conclude that these aerial insectivores are mistaking them for potential food. Yet I have also watched bats pursue flying tennis balls, hats, and Frisbees, and it is hard to conclude that an animal could mistake these objects for prey using an echolocation system capable of discriminating different insect species and even textures. Their behavior suggests at least curiosity, and perhaps play.

Some birds 'drop' *themselves* through the air. Various birds of prey perform an aerial ballet that starts when a pair approaches each other with talons outstretched. One of the birds flips upside-down to initiate the grasping of talons, then both birds spiral downward in a semi-controlled plummet, with wings outstretched. I once watched a pair of bald eagles perform this awe-inspiring feat in Ontario. Some biologists offer an aggressive interpretation, but the interaction has a distinctively playful – or at least thrilling – look about it. Perhaps it's a display of bravado – an avian variation on bungee-jumping, sans rope.

Similar shenanigans go on in the water. Hippos do backward somersaults. Dolphins leap two or three times their own body length above the ocean surface. Whales breach, heaving their huge bulk skyward before splashing down. This behavior might help dislodge skin parasites, but we don't know, nor do we know that it isn't fun into the bargain.

Other games

Animals don't play chess or volleyball, but they have their own games. Chimps finger wrestle, chase, and tickle. Gorillas do takedowns and throws. Elephants trunk-wrestle, and youngsters have a penchant for climbing up the backs of adults. Young Arabian babblers, who live in tightly-knit family groups, play games of king-of-the-hill, chase and tug-of-war. Seals body-surf. 'Peek-a-boo' appears to be a favorite game among African grey parrots. Kangaroos sometimes grab at falling leaves (a game I have enjoyed on windy autumn walks with my daughter). Spotted-necked otters go in for wrestling, tag, ducking, rolling, and friendly biting. These and other games are no doubt useful practice for developing and maintaining the strength and skills that living requires. When an anhinga juggles a stick in her beak, she mimics the art of orienting a captured fish just the right way to be swallowed without the spines sticking in her throat. It is also a sort of game, not critical to survival, and probably pleasurable.

Chasing is one of the simplest games, enjoyed by both humans and other animals. It is part of survival for many predators and their would-be quarry. It is therefore no coincidence that many animals, from gulls to gorillas, play at chasing. In the early morning of a March day, I watched for five minutes as three adult-sized (though perhaps juvenile) cottontail rabbits gamboled about a low-lying juniper shrub on a picturesque university campus in the American Midwest. They chased one another around, darting into or out of the shrub, running and leaping across the dewy grass. These are all useful escape skills. They are also clearly rewarding when there's no predator about, or else why would the animals expend valuable energy on them?

Play has been described in 27 species of kangaroos and their relatives, despite very limited information on the many small, rare or secretive species. Many studies are done in captivity, often in small enclosures where there may be less opportunity to express the full range of natural behaviors. Play-fighting is the most common form of play in kangaroos. Doria's tree-kangaroos play-fight lying on their sides. In red-necked wallabies, play-fights can last up to twelve minutes. Intense play-fighting may involve 'skipping,' in which partners jump around each other in a 'high-stance' posture. Adults self-handicap against smaller and weaker playmates. For example, rarely does the larger partner rear up or kick; littler playmates are also much more vigorous in their play-fighting.

Just as each animal is a unique individual, so too are the games that each devises to amuse him or herself. Mike Tomkies describes a game of 'footsie' he would regularly play with a semi-tame tawny owl, in which the bird would grab Tomkies' hooked forefinger in one foot, then hang on while Tomkies pulled his finger about. The owl, capable of mangling a finger in his sharp talons, was always gentle. Anthropologist and author Mary Thurston describes an adult male squirrel who makes daily visits to her back patio, where he pulls the towels out of the chairs and onto himself, hopping and twisting until he is completely covered. She doesn't have any proof that he's enjoying himself, but his behavior serves no other obvious purpose such as acquiring food, territory or a mate.

Captive dolphins produce bubble rings – hoop-shaped air bubbles that rise erratically towards the surface – from their blow-holes, which they flip or fuse, apparently for the heck of it. Dolphins in the wild muck about with seaweed – tossing, passing and balancing it, and using it for tug-of-war. Bottle-nosed dolphins often force Pacific white-sided dolphins away from prime surfing spots in the waves.

A variation on surfing is 'bow-riding': riding the pressure-wave at the front of a large moving object in the water, such as a motorboat. Bottlenose dolphins have also been seen riding the pressure waves of gray whales, humpback whales, and right whales. It could save energy in transit, but this is unlikely to be the sole motive, because the bow-rider can go only where the pressure-wave is going. That dolphins often leap and gambol and criss-cross during bow-riding further supports the notion that it is fun.

Powerful tides provide another potential for amusement. For years, medical illustrator Ken Heyman has watched laughing gulls riding the tides at an inlet in Ocean City, Maryland. The birds ride the tidal current as the tide recedes towards the sea.

When they get to the end and the current dissipates, they fly back and start again. Ken has seen no sign of foraging, and reports that the birds seem to do this whenever the tide goes out. There are several published scientific reports of birds riding water-flows. A 1934 report describes common eiders riding down rapids in southern Iceland, repeatedly dashing back to the same spot for another go. Adelie penguins bob along on tide-pulled ice floes and return to the starting point to do it all again.

On 13 June 1946, while watering his garden in Benicia, California, Emerson Stoner saw an adult female Anna's hummingbird doing something similar, as he wrote in the journal *Condor*:

> ... the water was flowing from the hose in a solid stream about three-quarters of an inch in diameter. [the hummingbird] flitted alongside the flowing stream and eyed it ... Then she took a position facing the stream, brought both feet forward and dipped them into the water. Finally she came at right angles to the flow and attempted to light on it as though it were a twig or limb and rode down the stream a way, repeating this stunt over and over again.

Nellie Tsipoura, who studied shorebirds for her PhD at Rutgers University in New Jersey and now works for the New Jersey Audubon Society, recalls seeing a different sort of water play one day while watching a flock of about 30 sanderlings in Texas. They all stood on a jetty, jumping as the waves broke over them. These gregarious little sandpipers would crane their necks waiting for the next wave to reach them, fly up and hover in the splash as the wave broke, and then land on the jetty again. Nellie is certain that they were playing, for they did this for the 20 minutes that she watched them, and they spent no time in any activity other than looking for the waves and jumping. She likened them to a group of children playing in the surf and jumping in delight. Nellie never tried to publish what she saw because she felt it would be not taken seriously. How many other intriguing animal behaviors are witnessed and forgotten because they don't meet mainstream scientific expectations?

Whereas once it might have been dismissed as the imaginings of scientists gone astray, there is now published evidence for play in reptiles, and even in at least one invertebrate group. Captive komodo dragons bang buckets, grab and shake shoes, and engage in tug-of-war over metal cans with familiar keepers; and according to ethologist Gordon Burghardt, the animals are neither hungry nor aggressive. 'Titillation courtship' has been noted for years in the young of certain freshwater turtles, but only recently has it been recognized as sharing many features with the social play of rodents.

Pigface, a fifty-year-old male Nile softshell turtle at the National Zoo in Washington, DC, during the 1980s and 1990s, often inflicted injuries on himself, probably because of his solitary, unstimulating existence. When new objects (basketball, hoop, sticks, etc.) were added to his enclosure, he promptly began to interact with them. Video analysis of Pigface's activity showed that he spent 31% of his time with his play objects – nosing, biting, grasping, chewing, pushing, pulling and shaking them. The hoop, which he sometimes swam through repeatedly, was his favorite. Providing new and diverse objects and rotating familiar ones greatly reduced the tendency of the turtle to become bored and return to self-destructive behavior.

Pigface is not unique among turtles. Similar object play has been seen in two Nile soft-shells at the Toronto Zoo. Researchers have also reported curiosity and object play in green and loggerhead turtles living in the wild, and in captivity. The sea turtles being kept at a rehabilitation center used to spend 100% of their waking time swimming in circles in their small (3 m diameter) tanks, raising their heads to breathe and occasionally changing directions. When square-shaped devices made of connected PVC pipes were affixed to the sides of their tanks, the animals immediately gravitated towards them. Their circular swimming dropped to about 30% of their time as they rubbed their sensitive upper and lower shells against the hard surface, swam through the opening, and rested on the bars.

There are scattered accounts of play in other reptiles, including terrestrial turtles. Burghardt also reports on a wood turtle who repeatedly climbed and slid down a board into the water, and a wild, juvenile alligator who 'stalked' a waterpipe dripping into a pond, snapped at the drops as they fell, and allowed them to spatter off his snout.

Recent studies demonstrate playful behavior in two species of octopus, the Pacific and the common. When seven captive common octopuses were presented with an opaque plastic bottle and a rectangular Lego® block by Michael Kuba and colleagues at the Konrad Lorenz Institute for Evolution and Cognition Research, in Austria, some showed possessive behavior towards the objects. Others displayed playful interactions, which included pulling or pushing the object towards or away, towing it, and passing it from one arm to another. The animals were more likely to play if they were recently fed, further suggesting that it was recreational. Jennifer Mather, a Canadian biologist who has studied octopuses for over 20 years, has seen an octopus repeatedly push an empty plastic pill bottle into the flow of water from an aquarium hose. When the bottle is propelled back, the octopus grasps it in a tentacle and places it in front of the hose again. Mather has also seen the creatures appear to play with bubbles.

Play can help relieve boredom. This may be especially significant for animals confined to the monotony of captivity. Captive elephants develop games to prolong their time outside, and captive orangutans are legendary for their escape attempts. One captive orangutan grabbed clumps of grass to make an insulating mitt, then climbed over the electrical hot wire surrounding his enclosure without harm. Captivity allows us to see first-hand what may occur relatively rarely in nature. It imposes on animals conditions that they may not normally encounter in the wild; it requires them to make the best of unnatural circumstances.

Deprived of their own kind to play with, they substitute with whomever is available.

Just as companion animals play with their guardians, play is not restricted to members of the same species. In 1994, a playful relationship developed between a Siberian husky and a polar bear near Churchill, Manitoba. On first meeting, the dog performed a play-bow, which the bear appeared to recognize, and the two tussled for several minutes, embracing and parting when the bear grew overheated. The relationship, which was documented by professional wildlife photographers and appeared as a cover story in *National Geographic* magazine, lasted a week until the winter ice formed and the bear left for his winter feeding grounds.

At the Karisoke Research Station in Rwanda where Dian Fossey's research team studied mountain gorillas, some of the younger apes would wrestle with a pair of golden Labradors who lived with the researchers. Some of these interactions were captured on film and are a joy to watch. Chimpanzees often initiate play with young baboons or humans. Hugo Van Lawick once watched five year-old chimp Frodo roaring with laughter as a baboon he was playing with tickled him. Baboons and chimps can be mortal enemies, but this interaction clearly indicates that the two species aren't always at loggerheads. On another occasion, Van Lawick watched two young bat-eared foxes play with an adult Thompson's gazelle who bucked and chased the young foxes in circles – bat-eared foxes are far too small to prey on gazelles.

A favorite game for elephant calves is chasing wildebeest. Two or three calves will charge, ears flapping, then watch their mock quarry scatter. Rosamund Young, a cattle farmer and an authority on cows, has watched calves playing 'tag' with a fox around a tree. The well-studied red-necked wallaby repeatedly approaches and follows Australian magpies (also noted for their playfulness), domestic cattle (usually calves) and insects.

Mischief

Some animal play has the smell of mischief about it. Among parrots, the kea of New Zealand is noted for vandalizing cars – removing chrome radio antennas is a favorite trick. Such is keas' propensity and aptitude for dismantling things that they have inspired a specific category of behavior: 'demolition play.' Van Lawick describes a banded mongoose leaping up and grabbing in its mouth the tail of a grazing male bushbuck in Tanzania. The bushbuck lowered his rump to the ground and the mongoose let go. The bushbuck soon resumed grazing but the mongoose had followed, and again leaped up and hung on to the bushbuck's tail for awhile. This time, the larger beast lowered his rump more slowly and continued to graze as though unconcerned. Van Lawick concluded that the bushbuck had experienced this 'game' before. In Austria, ravens have been spotted pecking the rear-ends of resting deer, and in Europe, little grebes sometimes tweak the tail feathers of swans, then quickly dive.

A variety of mischievous play goes on in cetaceans. Dolphins 'bop' floating seagulls or cormorants into the air, and harass puffer fish to provoke them into puffing up. Gulls may also be the butt of mischief by captive orcas, which sometimes bait them by letting dead fish float to the surface, then grab and dunk them before letting them go. These sorts of pranks suggest a sense of humor (see Chapter 9).

There are many accounts of peregrine falcons getting into mischief. For example, in Ontario, Alf Rider was one of several experienced birdwatchers who watched a pair of peregrines fly out over Lake Huron on several different days and simply harass the gulls there. The duo dive-bombed Bonapartes, ring-billed and herring gulls. Sometimes a herring gull would turn the tables and chase one of the falcons. There was no physical contact. Once, the two peregrines followed their sport by 'buzzing' a squadron of Tundra swans, flying within inches of their tails and also getting between them. The discomfited swans broke their formation and made a hasty landing on the lake.

The mobbing of birds by other birds likely has survival benefit, as when blue tits flutter angrily around crows who may be after their eggs or chicks. But some forms of mobbing seem to go beyond the call of duty. As a birdwatcher, I've learned that when I see or hear a noisy murder of crows milling about a large tree, there's a good chance there's an owl (or some other large raptor) there. Crows mob owls with enough zeal and vehemence to stir pity for the owl. Could there be some degree of entertainment in this for the crows, or are they just going overboard with their natural antipathy towards birds of prey? I have also seen a crow alight briefly on the back of a soaring turkey vulture, and a red-winged blackbird perching for several seconds on the back of a soaring red-tailed hawk. These interactions appeared neither friendly nor vicious. Might they be a form of mischief or play? I have not yet seen any descriptions of this 'aerial perching' behavior in the scientific literature. Perhaps the roots of such harassment lie in nest defense.

At least one account of behavior in an invertebrate cannot be discounted as a possible example of mischief. Jennifer Mather describes a red octopus aiming and squirting (with accuracy) jets of water at a visiting scientist and photographer, even though he/she never squirted at Mather or her assistant. A common octopus in her laboratory aimed jets at her or the observation stool she would sit on for two weeks after she had banged on the aquarium lid from which this particular octopus had been trying to escape. Whether these behaviors reflect positive or negative feelings in octopuses, they certainly reflect a capacity for individual recognition, and for fairly long-term memory.

One set of antics there can be little doubt about are those of Georgia, a captive chimp at the Yerkes Primate Research Center in Atlanta. Whenever she sees a visitor, she hurries to the tap to fill her mouth with water. She then casually returns to her routine activity, and will hold the water secretively for minutes if necessary before spraying it on the unexpected guest, to shrieks and laughs from her chimp-mates.

Thrill-seeking

Some mischief involves a degree of risk. There are evolutionary explanations for risk-taking in humans and other animals. It may impress members of the opposite sex or intimidate those of the same sex by showing courage and physical prowess. This may help explain why some people gravitate to high-risk forms of recreation, such as bungee-jumping, skydiving and mountain climbing. Many more of us derive exhilaration from less dangerous but similarly thrilling escapades such as roller-coasters and horror films.

For all its evolutionary swagger, the conscious experience of taking risks is thrilling. Few climbers scale mountains to improve their reproductive output. They do it for the physical and emotional rush it provides: the adventure, the challenge, the triumph over adversity, and the sense of accomplishment.

Do other animals thrill-seek? Is exhilaration on the agenda of pleasures in the animal kingdom? Orangutans in Tanjung Puting, Borneo, play a sport that human observers call 'snag riding,' which involves hanging onto a falling dead tree, then grabbing a vine or other vegetation to escape before the tree hits the ground.

My father remembers an impromptu game he played with piglets in a pen in New Zealand in the sixties. Always drawn to animals, he made a playful menacing gesture towards the piglets, who all ran out into the adjoining shed. Gradually, they crept back, their attention riveted only on dad, who stood among several other men. He repeated the gesture and it soon turned into a game, repeated several times.

My three rats engaged in a variation of this game. When I had them out of their cage, I draped a large old curtain remnant over their playing area. I would lure them with a clicking sound, and before long there was usually a moving form beneath the cover. When I playfully tickled them through the fabric, they would scurry away for shelter, but return within seconds for another go.

Some animals show ingenuity and boldness in pilfering food from unsuspecting humans. I recall vividly the occasion when a jungle crow snatched a biscuit from my hand as I walked away from a street vendor in southern India. I was on a brief stop during an eight-hour boat trip between two towns near the coast. The bird ambushed me from behind, and I was caught completely unaware. As I returned to the boat, I warned a fellow passenger to keep an eye out for these aerial pirates. She took this advice with bemused skepticism, then returned wide-eyed to the boat five minutes later to explain how she had lost half a sandwich to a black blur that had flashed past her face. Aerial piracy is routinely performed among birds, including gulls and jaegers. One group – the frigatebirds – is even named for this penchant. It is no simple art to strip a human of his food, and I suspect these crafty birds have learned their trade by watching others, as have the tourists who soon learn to shield their snacks with one hand. It undoubtedly also involves some degree of planning, for success hinges on an approach from the rear, and requires impeccable timing for an inattentive target and an accessible morsel. Bungled attempts could have grave consequences. It would be interesting to study how these birds would respond to different human subjects and food items in varying circumstances.

Ravens are inclined to goad dozing wolves, which suggests the exhilaration of danger given the risk of being caught and killed. Wolf authority David Mech and others have seen ravens dive at wolves resting on lake ice, walk up and peck sleepy wolves on their tails, and even alight on their backs. Wolves will lunge and stalk ravens, who may evade them often at the last minute, as in a game. Primatologist Frans de Waal describes how his tame jackdaw, Johan, who would fly just over the heads of dogs, taunting them into jumping and snapping after him. I have watched bold squirrels repeatedly approach to within an inch of the screen door at the rear of my home, behind which a cat sits riveted and ready to pounce. The squirrels seem fully aware that the cat is there, but are equally confident in their untouchability. The cat actually did pounce on one occasion, causing the squirrel to leap back and vanish in a flash. The saucy rodent returned within two minutes, as if to call on another adrenaline rush.

Similarly, the games we include in our relatively safe lives may have some origin in the desire for a sense of danger. Hide and seek, one of the most popular and timeless of children's games, is perhaps a modern reenactment of times when it was important to evade detection by hostile enemies. The same may be said of animals who must avoid detection by predators. I'm not suggesting here that it is 'fun' for a rabbit to be pursued by a fox, particularly if the fox has detected the rabbit and is giving chase. These times will be highly stressful for the rabbit (though surely exciting for the fox). But the state of being in an environment where danger may lurk is not conducive only to negative stress. I remember the feeling of excitement I had of wandering the wildlands of Kruger National Park in South Africa in 1985 as part of a scientific team studying the ecology of bats. I was not new to trekking through natural habitats, but this was the first time I had been out in a region where large, potentially dangerous animals (lions, leopards, hyenas, rhinos, buffalo and crocodiles) were fairly abundant. And while it is easy to sit here now in the safety of a study and reflect wistfully on the positive emotions of that time, it is fair to say that I felt intensely alive and satisfied with my existence. I recall a similar feeling when I set out alone at daybreak into a forest in western India known to be inhabited by leopards.

Or perhaps these are only the feelings of a pampered, urban human casting out into unfamiliar wild terrain. For now, we can only wonder, as Richard Adams does in *Watership Down*, his famous fictional tale of a community of rabbits:

> The sparrows in the ploughland were crouching in terror of the kestrel. But she has gone; and they fly pell-mell up

the hedgerow, frisking, chattering and perching where they will.

Conclusion

Play is a good jumping off point for a journey of discovery into the world of animal pleasure. It is widely accepted by scientists. We cannot experience directly an animal's feelings during play, but we can see the exuberant quality in it and compare it to our own play. Play is adaptive. It benefits animals by helping them develop and maintain physical and social skills and therefore to survive and to reproduce. But playing animals surely do not contemplate evolutionary benefits; their immediate motive to play is that it is fun.

Chapter 5
FOOD

The pleasures of sustenance

Most robins seem terribly glad to be eating worms.

Joseph Wood Krutch

For us, the pleasures of food begin with acquiring or preparing it. The sight, smell and even the feel of our food all contribute to our anticipation of it, which is itself a source of pleasure. We also enjoy the satisfaction of becoming sated. Of course, it is when we taste food that we experience the most obvious and familiar aspect of food pleasure. Sweet, salty, sour, and even bitter tastes can please us, and there are myriad variations on our experience of these flavors. Persimmons taste distinct from pears, raspberries from strawberries, kidney beans from pintos.

It doesn't end there. Food is an important part of human culture. Food preferences and our methods of preparing and presenting them vary diversely across ethnic and geographic landscapes. We are the only species to cook and to use recipe books. For us, food is inextricably tied to pleasurable sensations and experiences.

What about other animals – do they enjoy food? The role of the pleasures of taste in animals is almost wholly unknown. Yet there are innumerable clues that animals savor the flavor in their food. They show likes and dislikes, tastes that change through time, and anticipation of food. Some, such as language-trained apes and parrots, can even tell us their reactions to food. What's more, animals produce brain-rewarding opioids while searching for food and while eating.

There was a time when I would provide three ladies with breakfast in bed everyday. (I should explain that this trio were my rats.) They lived in a large cage I built for them, with platforms, sleeping box, exercise wheel, hammock, and other interesting things. Most evenings I let them out to explore an ever-changing landscape of blankets and boxes in our loft, or scamper about our rat-proofed bathroom, unraveling the toilet roll if I'd forgotten to put it away.

On the day I wrote this, their breakfast was a mixture of dried fruit (dates, banana, pineapple and papaya), fresh apple pieces, and – a rare treat – some pieces of fresh-baked peanut butter cookies. I mixed these morsels together in their ceramic food cup, and placed it on the shelf opposite their sleeping box.

Within a minute all three warm bodies left their slumber to investigate the morning's victuals. First Rachael, then her sister Veronica, then the somewhat older Lucy. Each one arrived at the cup in turn, peered in, sniffed intently, then pulled out a piece of cookie. I was not surprised at their response. They prized the occasional baked goods I gave them.

As they ought to. Peanut butter cookies are packed with energy and flavor. They contain more than double the caloric content of a comparable quantity of fruit. Evolution ought to favor an animal's preference for the most energetically rewarding food. A wild rat may face a varying and uncertain food supply, so it pays to consume the most bang-per-buck food when they have the chance. This probably explains our own penchant for fatty foods which now, in overabundance, do us more harm than good. By favoring the fat, we are responding to ancient survival pressures. Yet we don't do this consciously. We like French fries because they are delicious. Our brain and senses reward us for choosing them. Same for rats, who have been shown to work harder to get more palatable food offerings (more on this ahead).

We don't know what a rat feels like when she samples a peanut butter cookie. What we can be fairly sure of is that the sensation to Rachael is different from Lucy's, which is in turn not the same as Veronica's. Recent research indicates that how we see, smell and taste the world is as unique to the individual as are our genes. Sensory genes are not only vastly abundant, they are also highly variable. As a result, few individuals share the same set of sensory genes. A recent study of 189 ethnically diverse volunteers found that each had a unique combination of genes for smell. This may help explain why some of us adore Brussels sprouts while others cringe at the thought of them, why cats are finicky at the food bowl, and why, for example, pallid bats show individual preferences in the kinds of insect they eat.

Scientific interest in the sensory experiences of other animals is growing. Several recent studies have examined animal taste acuity. Hamadryas baboons are among the most sugar-sensitive of primates, supporting the idea that they use sweetness to choose food. A taste for sweetness is widespread in mammals, whose first experience with it is in their mother's milk. Either we acquire a taste for sweetness then, or we may be genetically predisposed to it.

Spider monkeys opt for higher energy foods given a choice. They can detect very low concentrations of sugar dissolved in water. Four species of monkey differed markedly in their acceptance of different concentrations of sour-tasting water. Olive baboons were the least picky, and squirrel monkeys the most. The latter preffered sweetness to sourness.

Many animals make positive and negative facial and gestural reactions to taste similar to those of humans. All monkeys and apes stick their tongue out repeatedly if offered something sweet. Rats, similarly, lick, draw their paws to their mouths and nod. Bitter tastes elicit a triangle-shaped mouth, chin down and treadling forelimbs. Studies of these animals' brains and nervous systems also suggest that these patterns derive from the same evolutionary source as for us.

Hummingbirds don't seem to smell to find food; the areas of their brains devoted to smell are poorly developed and hummingbird-pollinated flowers are typically scentless. Hummingbirds do, however, have a refined sense of taste and can detect differences in both the types and the concentrations of sugars in nectar. Like us, they prefer sucrose to simpler sugars such as glucose and fructose. Unlike us, they don't worry about tooth decay or fitness gyms.

In July 2003, I watched about 15 pigeons get into a feeding frenzy over half a peanut butter sandwich tossed to them near the entrance of a London art museum. The pigeons milled around, grabbing at the food and rapidly shaking their heads to break off chunks that could be swallowed, causing other chunks to fly up like popcorn over their heads. The pigeons selectively went for the sandwich filling, stooping their heads down to ground level and pecking sideways to remove these choice bits. It seems only reasonable to assume that they preferred the taste of the filling to that of the white bread. Sure, the peanut butter delivers more energy per peck. But pigeons do not consult nutrition charts. Evolution has endowed them with a taste for more sustaining foods that contain more fat and protein.

Evolving tastes

Few of us haven't experienced the disgusting smell or taste of food that has 'gone off.' If there are no visual clues that a food has become spoiled, then it is usually our nose that sends the message: 'Don't eat!' If that fails, our tongue provides the next line of defense. It is not only adaptive for you not to eat spoiled food, it is probably adaptive for the invading bacteria or other micro-organisms not to be ingested. Hence the microbe's warning system evolves alongside the forager's ability to detect it. In the case of poisonous foods, the penalties are even greater. Mistake thise deadly nightshade berries for edible elderberries and you might not live to reproduce. Little wonder that a poisonous plant usually has a bitter taste, at least to creatures for which it is poisonous.

Back at the pleasurable end of the taste spectrum, many plants produce fruits that contain delectable flavors. These treats are advertised by bright colors and alluring aromas. Fruits evolved to benefit the plants that produce them. The benefit is seed dispersal. Because plants are not mobile, they depend on other means – wind, adhesion to fur – to get their seeds away from the parent plant, which otherwise competes for the same resources. So intimate is this arrangement that, in some cases, specific plant and animal species have become almost exclusively codependent. The dawn bat is believed solely responsible for pollination of durian trees, whose stinky fruit is of high commercial value in Southeast Asia. The flower of the Brazil nut is

Figure 5.1 An Egyptian fruit bat savors a mango.

pollinated by just one species of bee, and the tree's nut depends for its germination on the agouti, a large South American rodent whose sharp teeth soften the seed coat. Other plants produce seeds that will not germinate unless they have first passed through the digestive tract of an animal. Such tightly knit plant–animal alliances are built on pleasure – specifically, animal taste preferences.

As Michael Pollan put it in *The Botany of Desire*:

> ... in this grand evolutionary bargain, animals with the strongest predilection for sweetness and plants offering the biggest, sweetest fruits prospered together and multiplied, evolving into other species we see, and are, today ... desire is built into the very nature and purpose of fruit.

Variety is the spice ...

Rats will enter a deadly cold room and navigate a maze to retrieve prized food (e.g. shortbread, meat paté, and Coca-Cola®). If they happen to find their regular commercial rat chow at the end, they quickly return to their cozy nests, where they stay for the remainder of the experiment. If they find a tasty treat, they eat it before returning home, then return repeatedly for more.

This is a rodent version of shunning the fruit bowl and dashing out for doughnuts on a rainy night. Lizards do it too. Big deal, you might say; the animals were simply trading off different needs. But variations on the same setup further revealed pleasure-based decision-making by the lizards. They made forays to the gourmet banquet only if the temperature in the cold corner was above a certain level. When it got too cold, they stayed in the warm and ate the nearby food. By varying the food offerings, it was discovered that the better the food in the cold corner, the lower the temperature the lizards were willing to tolerate.

Do animals get bored of the same old grub, as we do? I have often gazed at the graylag and Canada geese grazing on the lawns of my local park, wondering how their summer diet could be anything but tedious. Yet these same birds also receive frequent handouts from passers-by. In the wild, most grazing land presents a wide array of plants for geese, cows, and sheep to eat, and these animals are known to select from a variety of herbs, flowers, hedges, tree leaves and fruit. Cows will regularly go blackberry picking, and they eat plants, such as nettles, dock and thistles, that may have medicinal benefits. They also munch young hawthorn leaves and shoots, ash leaves, willow, wild thyme and sorrel. These foods provide further variety as they change seasonally from young shoots to mature flowering plants, then go to seed. Cow parsley, cowslip and cow-heat are all plants named for their association with bovine diets.

In *The Mind of the Raven*, Bernd Heinrich says of the raven chicks he has raised by hand-feeding that they often refused food if they were offered the same fare several days in a row. One can surmise purely adaptive explanations for this behavior, such as the importance of a balanced diet. Of course it is better to eat unvarying food than no food at all – the ravens refuse it because they can afford to. They are well-fed and regularly provisioned, and they know that someone is likely to ply them with something else – something different, for a change – when they get finicky.

My three rats have this same tendency to become bored. They are more likely to dive into a cup of peas or peel out of their nestbox for some banana slices if they haven't had these foods for a while than if they've had them the last three nights in a row. A similar preference for new flavors has been demonstrated in the lab. A 2003 study found that Norway rats and golden hamsters preferred novel foods following several days eating a single food. The researchers concluded that the animals were either avoiding becoming over-dependent on a potentially short-lived food source, or were averse to the risk of developing a micronutrient deficiency. A more tangible explanation is that the rodents tired of the same old dinner, and given the chance, they went for variety and tastiness. The rat/hamster study authors also saw preference for novelty only in a negative context: as an 'aversion' to familiar food. They might equally have interpreted the animals' preferences for the new taste according to a capacity for pleasure.

DeeAnn Draper, one of the human teachers of Koko, the famous gorilla taught to communicate with sign language at the California-based Koko Foundation, notes Koko's 'obvious and continuous demonstrations of gorilla pleasure at the anticipation or sight of food.' When asked to say a long sentence about lunch, Koko signed: 'Love lunch eat taste it meat.' That a gorilla would express a liking for meat also presents an interesting insight into the potential flexibility and changeability of an animal's individual tastes, and how exposure to unusual foods can lead to their being favored. In the wild, gorillas eat only plants and some insects. Humans aren't the only ones who can 'acquire a taste' for something.

Many factors in an animal's natural history influence his or her dietary selection, including seasonal availability of foods, nutritional requirements, hormonal changes and reproductive status. Beyond this, however, we may expect individuals to show a discriminating palate based on sheer enjoyment alone, particularly if food is abundant. In Monterey Bay, a film crew recorded a pod of orcas eating only the tongue of a gray whale calf they had systematically separated from his mother and killed. When naturalist Mike Tomkies found the tongue missing from a freshly fox-eaten deer carcass in the Scottish highlands, he interpreted it as a clear indication that tongue is a fox delicacy.

Individual tastes

An American acquaintance of mine who has shared her home for a decade with two cockatiels and a cockatoo knows their food fads inside out. These birds also know each other's preferences. In the spring, parrots become more attentive and affectionate towards their mates, searching for suitable nest sites and performing impressive songs and dances. In this case, Hollywood (a 15-year-old male cockatiel) is paired with Diego (a 13-year-old female cockatiel named prior to laying her first egg) and Maliboo (a 10-year-old rosebreasted cockatoo) with his human 'mate' Christi.

Like a pampered child, Diego loves apples, but not just any apple; it must be a Granny Smith, peeled and chopped. She also adores bananas, and will dance excitedly when she sees Christi peeling one. Hollywood, by contrast, won't eat banana at all and prefers only the skin of an apple. Of the three birds, only Diego has the sweet tooth (or, dare I say it, sweet beak). Hollywood and Maliboo eat raw carrots and broccoli, while Diego takes no interest in these foods unless they are slightly cooked. Hollywood prefers spinach and kale, Diego prefers kale and collard greens, and Maliboo prefers kale and chard.

Bernd Heinrich writes of presenting new foods to naïve young ravens. He found clear individual likes and dislikes. Some birds devoured pistachios with gusto, although they are challenging to learn to eat. Eggs were invariably appreciated. That the birds went enthusiastically for road-killed skunk, 'yanking fur out and poking into all orifices,' left Heinrich wondering if the ravens simply disregarded the smell, if they smelled at all, or if they like the smell of rotting skunk. When Heinrich buried a skunk carcass under snow the ravens couldn't find it, indicating that they use landmarks, and not scent, to locate food.

Like us, some animals like to combine food, apparently to enhance the taste experience. Maliboo, Christi's rosebreasted cockatoo, takes a piece of shelled pecan and eats it into the exposed flesh of a mango. He also occasionally gets a treat of cheese, which he combines with opened cherry tomatoes, alternating between nibbles of the cheese and the tomato seeds. Five thousand miles away in Rwanda, eco-tourists may watch a mountain gorilla tear a bamboo shoot from the ground, split it with her teeth, then stuff it with *Droketsa* leaves before passing it to her daughter to eat. It's a gorilla's sandwich recipe. A captive Indian elephant named Siri, confined in a small zoo exhibit, was often seen to step delicately on an apple or orange, split it open, then rub it on her hay, a behavior that her keeper took to mean that she was adding flavoring to her food. When an automated shower was installed at the Columbus Zoo that the elephants could turn on and off for themselves, one of several innovative uses for it was to dampen hay, which elephants prefer to dry hay. I recently watched an eight-year-old captive sulfur-crested cockatoo dunk a digestive biscuit into a hot cup of tea then scrape off the softened outer layer.

Joanna Burger describes in engaging detail the gustatory habits of Tiko, her adopted parrot. Tiko the parrot's responses to food are among the many revealing aspects of his life. In the second year after Burger acquired him, Tiko informed Joanna that he was ready to expand his diet by flying down to their dining table one morning, stripping a piece of cream cheese from a bagel and downing it quickly. From that point on, he became something of a gourmand, eating most everything Burger and her husband did, including pasta, pudding, corn on the cob, sweet potatoes and ethnic dishes. Chinese take-out was a particular favorite, especially the snow peas, peanuts and baby corn. Tiko also developed a taste for chicken, especially the marrow, which he would extract with his tongue after splitting open the bone with his jaws. Nuts and chocolate would bring on cheeping, trilling and squawking. Refused extras, he would sulk. Another of Tiko's favorites is cheese, and Burger describes how he would become 'testy' and 'peevish' when it was rationed.

The pleasure of the palate may be much broader in most animals than commonly believed. In the wild, diet is often limited more by opportunity than preference. Scavenging for meat or bones is not uncommon for otherwise vegetarian animals. Similarly, carnivores will eat vegetables. Most who have lived with dogs will know that they will sample widely from cooked vegetables offered them. Even cats, among the most purely carnivorous of mammals, will eat vegetables. I've lived with cats who nibble at cantaloupe, papaya, hummus, margarine and soymilk. One would even lick the juice from a freshly cut orange.

Where there is preference for some type of food, it suggests that there is pleasure associated with it. While my vet can give me reasons why dry food is healthier for my cats, my cats give me clear signals that they prefer the wet stuff. Many is the time when I've seen a cat meow in anticipation of, and run for, some wet food that I am preparing to add to a bowl that already contains dry food.

Come and get it – the significance of anticipation

If animals can anticipate a future luxury or ruminate on past pleasures, then their lives hold a greater capacity for pleasure than if they are only occupied with the immediate present.

For humans, at least, we know that much of pleasure is the anticipation of it. Wanting to eat, based on appetite or craving, and liking to eat may involve separate parts of the human brain.

There is plenty of evidence that animals anticipate pleasurable events, most obviously food. Ivan Pavlov's famous studies on conditioning with dogs used drooling as an indicator that the animals were anticipating food. Bernd Heinrich describes the saliva dripping from the bills of a group of semi-wild ravens when he empties a bag of cheese puffs into their enclosure.

There are countless examples of captive animals working for food even when nutritionally adequate fare is readily available to them in a nearby dish. Pigs will root, rats will pull a lever, and pigeons in Skinner boxes will peck at a disk to obtain food that is also freely available. This phenomenon, called 'contra-free-loading,' may result from the urge to search for food, which is an understandably critical behavior for survival in the wild. It also suggests that the anticipation of feeding is rewarding in itself. Further, it demonstrates the motivation shown by animals kept in unstimulating cages to be engaged in doing something, and to exert some control over their lives.

Anticipation is central to the matter of animal pleasure. Anticipatory behavior is further evidence that animal experience is not limited to the moment, and that they can be affected by perceived future events. This ability to relate to things not in the immediate suggests that the past may also be a source of pleasure – or displeasure – as it can be for us.

In the 1920s, Yale University psychologist Otto Tinklepaugh conducted a celebrated experiment that illustrated expectation of a favored food in two monkeys – a female cynomolgous macaque named Psyche, and Cupid, a male rhesus macaque. Each monkey learned to find a banana hidden in a particular outdoor spot. When Tinklepaugh secretly replaced the banana with a lettuce leaf, the monkeys' behavior was just what we would expect of an expectant individual. At first, the lettuce was left untouched, and the monkey looked around, inspecting the location and its surroundings over and over. In some cases the monkey even turned to the experimenter and shrieked at him. Only after a long delay would the monkeys content themselves with the lettuce. As Frans de Waal said of these experiments: 'How to explain such behavior except as the product of a mismatch between reality and expectation?'

On the other side of the coin there is satiety. It's a feeling we know as subtly pleasant – unless we overindulged. We might expect that other animals feel better after a meal, but detecting that is another matter. Recently, a team at Cambridge University discovered that sheep can discriminate between the face of a contented sheep and a stressed one. Presented with two doors they could push open to gain food, only rarely would sheep push the door on which was pictured a very hungry sheep. They vastly preferred the door with a picture of a sheep who had just had a meal. Incidentally, the sheep also strongly favored a door with a smiling human face over one with an angry human.

These findings hint that sheep can read emotions from faces. They also suggest that sheep (and, I suspect, goats, pigs, cows, gazelles, gnus and other gregarious ungulates) might read the mental state of one of their own kind. It would be interesting to see the results of an experiment on how well humans could discriminate the photos of contented versus stressed sheep.

If sheep faces can convey emotions, it seems likely that the faces of other animals – particularly group-living ones – can too. The Cambridge team's method could be applied to many other species to test this. Of course, animals' feelings may be expressed in myriad ways. I know when my rats are playful by their movements, not by their facial expressions, which remain unchanged as far as I can tell. Rats also express joy with short calls in the region of 50 kHz (see Chapter 7). Being well into the ultrasonic range, we cannot hear these sounds unaided; other rats can.

Indeed, the privacy of many of the shades and nuances of pleasure is one of the reasons it has been so little studied. Too often, an absence of overt expression of pleasure is taken to mean absence of pleasure itself. It's a short-sighted view. Watch someone eating. Especially if they are alone, you won't find facial expressions or other conclusive signs that the food is pleasurable. Yet, eating is among life's most basic pleasures! For us, the feeling of tasting, chewing and swallowing food is intensely rewarding and satisfying, even when we are not hungry, as when we tuck into a dessert after the main course.

Tasteful fish

Might fish enjoy their food? They can certainly taste. Studies conducted in the 1920s found that minnows could be trained to tell apart sweet, salty, sour and bitter substances, and several natural sugars. A study published a few years later showed that hungry minnows can detect sugar at concentrations 512 times lower than can humans, and salt at thresholds 184 times lower. The removal of olfactory tissue from these unfortunate fishes' brains did not alter these thresholds, leading the authors to conclude that the fish had a true 'gustatory sense.'

In June 2001, I played a simple trick on a small group of koi (brightly-colored members of the goldfish family) in a small fishpond in an in-law's backyard in Michigan. I held my hand over the pond and made movements as if sprinkling food onto the water. The fish came rapidly swimming to the place beneath my hand, where they loitered briefly before dispersing again. These fish are fed daily by someone who holds a hand out over the pond to sprinkle in fish food. Their behavior of orienting towards a proffered hand will be familiar to practically anyone who lives with aquarium fish.

What does it mean when fish race to a food source? The traditional explanation for this behavior is that the fish have been conditioned to associate the appearance of a human hand over their tank with the arrival of food, because one is routinely followed by the other. Presumably, they rush to the site because an early arrival is more likely to be rewarded with food. Fellow fish are also competitors in the pursuit of finite resources.

It could also mean the fish enjoy the food, and that taste makes haste. This interpretation doesn't exclude competition as a basis for their haste. Competition and pleasure could well be at work if our own experience is any guide. Consider your own motives in situations where food is shared among a group, such as communal pizza at a party. Many of us feel a sense of urgency about eating our first piece to improve our chances of having a second. Even in well-fed societies, where having enough food is among the least of our worries, an ancient urge to get one's share is present, and we often eat faster and ingest more than we ought at a buffet. But the conscious experience of 'pizza pressure' is not subsistence, but the pleasure of eating an extra piece.

To date, most research on fish has examined their physiology, not their experiences. What is known is consistent with the possibility that they get a kick from food. Fish brains have well-developed olfactory bulbs, and fish respond keenly to odor in water. Fish also have an abundance of taste buds, and nerves that conduct signals from these to the brain. As early as 1903, taste buds were known to occur in at least 35 fish species, in their mouths and elsewhere. The entire surface of a carp's mouth is peppered with tastebuds that enable it to sort food from detritus. The entire body of the bullhead (a kind of catfish) is coated with taste sensors. It's a useful adaptation for a creature which forages – as many fish do – in murky water.

Since fish respond preferentially to the smells and tastes of different substances, since they have the sensory and neural anatomy to sense what they're doing, and since they have evolutionary parallels with other vertebrates, it is not unreasonable to expect that they derive some sort of reward from the act of feeding.

Conclusion

A connection between food and pleasure should be strong and widespread in the animal kingdom. Food is fundamental to survival and pleasure is a strong motivator. Furthermore, it is vitally important to survival that one have a keen sense of taste for avoiding toxic or unsuitable food.

Evolution and the experience of pleasure work in tandem to promote survival. This applies to food as it does to play, sex and any other important survival or reproductive activity in the feeling organism. As it applies to monkeys and minnows, so too rats and lizards. My rat Rachael and her pals may have brains and senses 'wired' to dive for the peanut butter cookie first. But hey do so not merely to obey some ancient calorie-loading instinct. They grab the cookie because it tastes good.

Chapter 6

SEX

Procreation with recreation

There has been such a lengthy and deafening silence on this enormous subject, the sexual behavior of animals, that the first revolutionary survey of the scientific literature had to be over 700 pages long!

Ralph Abraham, in the frontispiece to Bruce Bagemihl's book *Biological Exuberance*, published in 1999

Warning: making the case for sexual pleasure in animals requires venturing into territory that may be distasteful to some readers. If you may be one such, I suggest you skip to the next chapter.

It is hard to overestimate the importance of reproduction to an organism. Without it, species would cease to exist. While not all organisms reproduce sexually (bacteria, for example, simply divide in two), in the vast majority of vertebrates and invertebrates females and males unite bodily (and bawdily) to procreate. Because furthering the line is so important, natural selection strongly favors behaviors and experiences that promote finding mates, mating and, where necessary, raising young. Among sexually reproducing animals with nervous and sensory systems, sex should be neither neutral nor nasty, but nice. Procreation and recreation make good bedmates, as it were.

Until recently, what was known of sexuality in the animal kingdom was confined mostly to the pages of academic journals and textbooks, the unpublished field notes of ethologists, and the memory banks of unwitting onlookers. Then, in 1999, a book was published to change all that. *Biological Exuberance: Animal Homosexuality and Natural Diversity*, by Bruce Bagemihl, a Seattle-based biologist and author, is a 750-page in-depth survey of homosexuality and other forms of non-procreative sexual behavior among mammals and birds (Bagemihl omitted other animal groups for want of time and space).

To peruse Bagemihl's masterpiece is to meet evidence of pleasure-generating sexual behavior on practically every page. In addition to detailed citations of source material, *Biological Exuberance* is liberally sprinkled with photos and illustrations: two young bonobos engage in an open-mouthed kiss, a male giraffe mounts another, a walrus stimulates his erect penis with a flipper, an orangutan masturbates with a tool she has fashioned from a piece of liana, an adult male bonnet macaque fondles the scrotum of another, two manatees rub their penises together then embrace in a '69' position with each's penis in the other's mouth, a young red fox female mounts her mother, and a white-tailed deer arcs his back to rub his erect penis against his ribcage. If this material had been in print in the 17th century René Descartes might have abandoned his nonsense of animals as unfeeling brutes, saving a lot of pain and suffering.

Just for the fun of it?

Sexual activity in animals is conventionally portrayed as all business and no pleasure. 'We know next to nothing about the evolution of sexual pleasure,' says animal 'sexpert' Olivia Judson, author of *Doctor Tatiana's Sex Advice for All Creation*, and this is surely part of the problem. From textbooks to television programs on animal behavior, the message is that practically everything that animals do is for an evolutionary purpose, sex being no exception. The participants are responding to powerful urges over which they have little control. The solemn purpose of this behavior is the production of offspring. The idea that animals may be enjoying themselves is not explicitly rejected. It isn't mentioned at all.

Against this backdrop, we would not expect animals to waste precious time and energy on sexual activity when there is no hope of producing viable offspring.

Yet they do.

Many animals routinely copulate or engage in other sexual activities outside of the breeding season, including during pregnancy, menstruation (in mammals) and egg incubation. Such non-procreative activity may even constitute a large proportion of the animals' sexual behavior, as it does, for instance, for common murres, proboscis monkeys, addax antelopes, rhesus macaques, wildebeest, golden lion tamarins and mountain goats. Another variation on the theme of wasteful sex is group sexual activity wherein few if any participants are passing along genes. These sorts of orgies have been spied in spinner dolphins, gray and bowhead whales, swallows and herons.

In his exhaustive survey, Bagemihl presents a veritable *Kama Sutra* of variations on the theme of sexual activity that has no chance of producing offspring. He discusses several forms and many examples of non-copulatory mounting, including: mounts without erection, mounts with erection (but with no penetration), and reverse mounting in which a female mounts a male (usually without genital contact). Another variation is mounting from the side or in positions from which penetration is impossible, which has been reported in mammals such as Japanese macaques, waterbuck, mountain sheep, takhi (Przewalski's horse), collared peccaries, warthogs and koalas, and birds including ruffs, hammerheads and chaffinches. Animals also go in for various forms of oral sex, stimulation of genitals using hands, paws, or flippers, and anal stimulation. Even interspecies sexual coupling is not unheard of.

Skeptics might say that these misguided antics are the fumblings of young, confused or otherwise inexperienced animals, though cases of oral or manual stimulation and cross-species interactions strain such claims. Behaviors that probably most suggest that sexual stimulation is pleasurable are variations on auto-erotic behavior. Given its utter futility with respect to procreation, masturbation suggests that the performer merely seeks pleasure. It is widespread in mammals, and practiced about equally by both females and males. It is known from at least seven mammalian orders, including primates, carnivores, bats, walruses, hoofed mammals, cetaceans and rodents. Masturbation appears less common in birds, but is by no means absent.

What of homosexual behavior? Most biologists today recognize same-sex sexual interactions as being part of the normal, routine behavioral repertoire of the animals who engage in it. Currently, at least 300 species of vertebrates are known to practice homosexuality. In most cases, participating individuals will also engage in heterosexual behavior, and their sexual life history can most accurately be categorized as bisexual. Homosexual behavior is clearly a bad idea from a strictly procreational standpoint. Practitioners may benefit by gaining practice, releasing sexual or social tension, or winning food. Nonetheless, the accompanying pleasure is probably what makes it such an effective social lubricant.

My three female rats regularly interacted sexually, especially during evening frolics outside their home cage. These interactions typically entailed chasing one another and then mounting, arching their backs and raising their heads and rumps into the characteristic 'lordosis' rigid U-shaped posture that facilitates coitus, plus some boisterous sniffing and licking of one another's vulvas. Usually only one rat was the focus of sexual attention, probably because she was in oestrus. Her demeanor and movements were friskier than the others. If I tickled her playfully at the nape, she would often assume the lordosis posture, then after a few seconds dart off again.

In birds, homosexual activity can extend to nesting and parenting. It is not uncommon among parrots and gulls for females to shack up together and go through the rituals of courtship and nesting. In New Zealand gulls, females often mount each other and males are relegated to little more than sperm donors. Subsequently, the females build nests, incubate in turns, and raise one or more chicks. Female-biased sex ratios are thought to be the cause of these unorthodox nuptial arrangements, though this explanation is unsatisfying, for clearly at least one of the paired females has mated with a male. In other species, such as mute swans, male pairings have occurred; such couplings are manifestly unproductive for siring offspring.

Figure 6.1 Red kangaroos enjoying sex.

The elusive clitoris

The existence of an organ whose sole function is to give pleasure would be convincing evidence that at least some animals experience sexual delight. Enter the clitoris?

The clitoris occurs in all female mammals. But, as Bruce Bagemihl laments, this naughty knob of tissue has been the victim of embarrassed silence throughout much of scientific history. His survey of the *Zoological Record* (1978–1997) – which contains over a million documents from over 6,000 scientific journals – found 539 articles dealing with the penis, and only 7 on the clitoris. Slightly less discouragingly, when I put 'penis,' then 'clitoris,' into the online journal database PubMed in mid-2005, I got 29,107 and 1,303 hits, respectively.

We can never know for sure whether or not the clitoris generates nice feelings in non-human mammals. That it does in women is one reason to favor the possibility. At least one leading thinker in the field – American anthropologist Sarah Hrdy – believes that females were shuddering in ecstasy long before women walked the earth. One of Hrdy's theories for the evolution of the female orgasm is that it aids sperm transport by providing muscle contractions during intercourse. If so, then there is every reason to expect the phenomenon to be widespread in the animal kingdom, where procreation is the evolutionary, if not the everyday, goal. On the other hand, perhaps there is no evolutionary basis for female orgasms at all. In *The Case of the Female Orgasm*, philosopher of biology Elizabeth Lloyd dismisses 20 adaptive theories for female orgasms, and concludes that they are a happy accident, based on the sharing of early embryonic tissue by the male and female genitalia.

Most of the behavioral evidence for female orgasms in other animals comes from observations of other primates. Female stump-tailed macaques appear to experience orgasm, as do female bonobos, siamangs and marmosets.

Might the clitoris be a spandrel? This lovely term (borrowed from architecture where it describes the roughly triangular wall space between two adjacent arches) was coined by evolutionary biologists Stephen Jay Gould and Richard Lewontin to denote a feature that exists more by happenstance than because of any direct evolutionary benefit. Male nipples are candidate spandrels because they appear to have no use in survival and reproduction. They probably persist in males simply because many of the genes that code for them in females are also possessed by males. Like male and female nipples, penises and clitorises share genetic origins. The clitoris is the female developmental counterpart to the penis. It has a shaft and a tip – the glans – just as a penis does. Like the penis it is also packed with sensitive nerves, and becomes engorged with blood when stimulated. Early in fetal development the external genitalia of males and females are almost identical, and it is only later that the penis and clitoris become clearly distinguishable.

Thus, the case might be made that the clitoris exists merely as an evolutionary tag-along – or spandrel – of the penis, and that, like male nipples, it has no important function.

If the clitoris were a spandrel organ, we might expect it to gradually lose those features that give penises their functional significance. A vestigial, useless penis ought to become small, soft and unresponsive to touch. The mammalian clitoris is definitely erectile. It is certainly sensitive for women, and appears so in those mammals for which data exist. And while it is externally usually much smaller than a penis, belated scientific study is revealing that – in women at least – most of its mass is internalized. A woman's clitoris measures nearly 10 cm and extends snugly along the vagina. An elephant's clitoris is over 40 cm long when erect and engorged, while that of the bizarrely equipped female spotted hyena looks identical to the male's penis, featuring a long and pendulous clitoris, accompanied by two pseudotesticles containing fatty tissue.

Despite science's general reluctance to explore animals' experiences of sex, they are nevertheless used as experimental 'models' to try to shed light on human sexual function and dysfunction.

Researchers at McGill University found that blood flow to the clitoris and vagina of rats increases when they are sexually excited. The researchers stimulated the rats' clitoral and pelvic nerves with electric currents and measured blood flow with lasers. At Boston University School of Medicine researchers have been studying clitoral and vaginal responses to various drugs in rats and rabbits, again with the aim of better understanding human sexual response. They report the same sorts of tissue behaviors experienced by women, including rhythmic relaxation and contraction of vaginal muscles, increases in vaginal wall pressure and vaginal blood flow, and increases in genital blood flow and lubrication. If sex is pleasurable, then we should expect these sorts of physiological responses.

Lascivious life

Theorizing aside, the best evidence that animals enjoy sex is their behavior. So let's take a brief tour of the salacious shenanigans of the animal kingdom/queendom, starting with our closest relatives then branching out to some lesser-known examples.

Primates: bonobos and chimpanzees

The Primate order includes our closest relatives, the Great Apes. Two of these – bonobos and common chimpanzees – are noteworthy for their prominent and diverse sexual behavior.

More slender and graceful than their common chimp cousins, bonobos are confined to the lowland forests of the African Congo River Basin. Their more human-like appearance is enhanced by their greater tendency to walk on two limbs. Bonobos live in small communities numbering up to about sixty. Their diet is primarily fruits and leaves, and piths from plant stems, supplemented occasionally by insects, small mammals, and aquatic animals.

Bonobos are highly social. They have evolved many ritualized social behaviors that may help maintain lower levels of social conflict than are found in common chimps. Sexuality is prominent in bonobo society, and female–female bonds especially. Females are in oestrus or false oestrus for nearly half the time. They have a very large clitoris that is more externalized than that of humans and therefore more accessible for pleasurable interactions. Females engage in a unique form of mutual genital stimulation, termed genito–genital rubbing, or 'gg-rubbing,' usually performed in a face-to-face embrace, one female on all fours over the other, whose legs are wrapped around her partner from below. The pair stimulate each other's clitoris by rapidly rubbing their genitals together in opposite directions from side to side, about twice per second in a coordinated effort that maximizes mutual friction. A single gg-rubbing session typically lasts about 15 seconds (a typical copulation lasts 12), and may be repeated several times in succession. Grinning, grimacing, and screaming accompany most of these interactions, as does genital engorgement, with both glans and shaft of the clitoris becoming fully erect. These behavioral and physiological features of gg-rubbing indicate intense pleasure, and probably orgasm. In a typical bonobo colony, females usually have several same-sex partners, and gg-rubbing may sometimes involve three or more females at once. On average, a female engages in a gg-rubbing session about once every two hours.

Male bonobos also go in for various homosexual activities, including 'penis fencing' from various positions, rump rubbing, fellatio, and manual penile stimulation. Male and female bonobos also masturbate solo, and males especially will sometimes use inanimate objects to stimulate themselves. In all, some 40–50% of all bonobo sexual interactions are homosexual with two-thirds to three-quarters being female–female.

In sum, bonobos are highly sexual creatures, and most of it has nothing to do with making baby bonobos.

Common chimpanzees (henceforth: chimps) live in 'fusion–fission' communities. Small groups of typically five or six individuals consort with each other within a larger community, or 'troop,' of up to 60 or more. All troop members recognize and know all others, and membership in the smaller groups changes regularly. Chimps are promiscuous, with females mating with numerous males. The promiscuous mating system of chimps is believed responsible for the evolution of chimps' large testicles that produce large amounts of sperm. Promiscuity selects for 'sperm competition' because if one male cannot prevent another from mating with a female, then the best chance that his sperm will fertilize the female's egg is to deliver as much sperm as possible, thereby increasing his chance of winning the sperm lottery.

In addition, female chimps, like bonobos, do gg-rubbing and mountings. They also stimulate the other's genitalia orally. Males have manual, oral, genital and anal contact with other males. These activities often occur in social contexts such as greeting, enlisting support, reconciliation or reassurance. They are also often combined with affectionate gestures such as embracing, kissing, grooming or nuzzling.

Non-reproductive heterosexual liaisons are also common in chimps. These include cunnilingus, fellatio, manual stimulation, 'bump-rumping,' masturbating alone or together, mounting without penetration and copulation during much or most of pregnancy. Male chimps have also been seen copulating with female savannah baboons in the wild, a union that obviously has no potential for offspring.

Masturbation is also common (notoriously so among all captive apes), and in males may take the form of auto-fellatio. In his book *Chimpanzee Politics*, Frans de Waal describes the masturbatory sessions of a large female, named Puist, in the colony he studied for several years at Arnhem Zoo in The Netherlands. These sessions lasted about one minute, during which time she made rapid finger movements through her vulva. De Waal noted no apparent change in Puist's expression during these bouts, but he wrote: 'it must have pleasant effects, otherwise why would she do it?' De Waal also noted copulating female chimps to produce a high single scream at the point of climax.

Other primates

Male white-handed gibbons often initiate sex with others, rubbing their erect penises together, which often results in orgasm. This typically occurs between father–son pairs, who face each other during the exchange, unlike heterosexual copulation, which is typically performed front-to-back. Males of another gibbon species, the siamang, practice a similar form of genital rubbing, and occasionally perform fellatio. Oral and manual genital stimulation is commonly seen in heterosexual pairs in both species.

Female hanuman langurs mount one another, grunting and grimacing. Male hanuman and nilgiri langurs do too, accompanied by a number of affectionate activities typically associated with sexual arousal. Grooming is intensely pleasurable, and may even lead to ejaculation. Japanese macaques have long tails, which females often use to rub their clitorises. Captive male rhesus macaques sometimes have anal sex rather than sex with a female. In his survey of primate sexuality, Alan Dixson lists published accounts of masturbation in 17 species of wild or semi-free-ranging primates.

Primatologist Suzanne Chevalier-Skolnikoff described the 'climax face' of female stumptail macaques during orgasm. Female monkeys exhibit physiological responses during copulation that are similar, if not identical to those of women. These include rhythmic contractions of the vagina, pelvis, and/or uterus, increases in breathing and heart rate, clitoral engorgement, and vaginal expansion.

These and related phenomena are known from at least twenty other (non-great-ape) primate species. And it is likely that the lack of information on the remaining 180 or so primate species is a feature not of its absence, but of our ignorance.

Seals and manatees

Gray, northern elephant, and harbour seals partake in a variety of non-procreative heterosexual activities. Near-term pregnant gray seals, for example, often copulate after coming ashore. As with many mammals, the testes of gray seal males are seasonally inactive and fertilization is impossible. Yet they will still copulate with females during this period. Heterosexual matings occur between gray and harbour seals, and these, too, are of course not going to result in young.

In some species, such as New Zealand and Australian sea lions, many males have virtually no access to females because they are too young, small, and/or inexperienced to compete with dominant males. Even taking breeding males into account, scientists have estimated that the average male copulates with a female only three to four times in his life, and many males never do so. The compensate with lots of homosexual activity. This behavior may have multiple benefits: practice for the real thing, alleviating frustration, and fun.

Prolonged abstinence is also typical for walrus males, who spend much of the year segregated from females while the latter attend to childbearing and rearing. Homosexual coupling occurs about five times an hour, especially in the water. Males also stimulate their arm-long penises with their flippers.

Male West Indian manatees interlock themselves in a variety of ways, each distinct from the heterosexual mating position. They sometimes embrace head-to-tail, or crosswise, and mouth and caress each other's bodies, nibbling or nuzzling the genital region. Males so engaged sometimes utter high-pitched squeaks, chirp-squeaks or snort-chirps. They may cavort for hours in groups of up to four.

Carnivores

Lionesses may lick their lions' genitals as a prelude to copulation. Cheetahs of both sexes lick their partner's genitals during heterosexual courtship. Male lions in captivity have been seen rolling their hindquarters forward over their heads and stimulating their penises with a foreleg.

Dogs, being outgoing, are prone to putting their sexual proclivities on display. Many readers will have suffered the indignity and embarrassment of their leg being made the object of amorous advances by a dog.

Cetaceans: bottlenose dolphins and spinner dolphins

Bottlenose dolphins range widely across the world's oceans. They are gregarious, and may form large pods of hundreds of individuals. Within pods, they tend to form smaller social units: mother–calf pairs, juveniles, female bands or adult males. Calves typically stay with their mother for three to six years, depending on when she becomes pregnant again. Bottlenose dolphins are social and affectionate. Sexual interactions occur year-round. Touch is important, and rubbing, caressing, mouthing and nuzzling are prominent parts of their courtship and sexual activities.

Males and females have a genital slit, so penetration is possible in both sexes, and the penis, the tip of the nose (the beak), lower jaw, dorsal or pectoral fin, or tail fluke are all pressed into service. Female spinner dolphins may 'ride tandem' on each other's dorsal fin, the female beneath inserting her fin into the genital slit of the other, and the two swimming together in this position. 'Beak–genital propulsion' also occurs, as does anal penetration.

Masturbation is also common. Males and females often rub genitals against each other or other objects, and this may escalate into a playful game. In her book *Dolphin Chronicles*, Carol Howard describes how a hurdle apparatus set up to exercise captive bottlenose dolphins prior to release became a sex toy. One male, Echo, began deliberately failing to clear the bar, instead rubbing his genitals against it with each pass. The behavior caught on and the captive pod was given a 'rub line' for this purpose. Bottlenose dolphins will 'mate' with sharks and even sea turtles, inserting their hooked penises into the soft tissues at the back of the turtle's shell.

Other dolphin species are highly social and sexual. The term 'wuzzles' has been coined to define groups of a dozen or more spinner dolphins of both sexes in an orgy of sexual behavior. Female spinner dolphins often appear pleasure-oriented, with vaginal penetration and insemination playing second fiddle to clitoral stimulation.

Dolphins can make strong enough sounds to stimulate each other at close range. Spotted dolphins perform 'genital buzzing,' in which an adult directs a rapid stream of low-pitched clicks at the genital area of another, usually a calf. Genital buzzing usually occurs between males, but is also a component of heterosexual courtship in this species.

Living underwater and relatively hidden from human onlookers, the comings and going of larger whales and their kin are something of a mystery. There are good reasons to expect that these animals enjoy sexual contact. They are big-brained, intelligent and long-lived. Many are highly social, traveling in groups, or pods. Needless to say, they are also well endowed: aroused males sport penises of 1.8 meters or more, and females are obviously able to accommodate them. Because of their shape and size, the case can be made that tactile sensitivity in the genital regions of whales aids the achievement of genital contact and penetration during copulation. Genital areas of both male and female orcas, for instance, are richly served with nerve endings, as they presumably are in the larger whale species.

Hoofed mammals

Female mule deer sometimes mount each other when they are 'in heat.' Males, who are infertile during the non-rutting season, nevertheless frequently mount females with erect penises, even though this rarely involves penetration. These mounts may last 15 seconds and occur 5 to 40 or more times with a single doe. Bucks of both mule and white-tailed deer also unsheath their penises, lick them, then rub them against their ribcages with pelvic thrusts and rotations until they ejaculate. For male red deer, wapiti, moose and caribou, sensitive antlers serve as erogenous zones, and when rubbed against vegetation may stimulate erection, and ejaculation.

In species where large, dominant males may monopolize females, mating and reproducing are simply not options for many males. It makes sense that those left on the sidelines do homosexual or autoerotic things for at least two reasons: (1) it gives them experience with courting, mounting and other sexual behaviors that they may be able to put to good use later, and (2) it keeps their plumbing in good working order should an unforeseen opportunity arise. From a strictly evolutionary perspective, if these behaviors confer some eventual reproductive benefits, then they should persist. If sexual activity is pleasurable it will reinforce these behaviors by motivating and rewarding them.

Male giraffes neck. Two individuals rub and entwine their long necks, often leading to both males' penises becoming erect. After 15 minutes or so, one male may suddenly stiffen, his neck held forward in a posture thought to indicate intense sexual excitement. Necking males also commonly mount one another and are believed to reach orgasm based on observations of fluid, presumably semen, streaming from their erect penises. Because of the 15 month gestation in giraffes, and the minimum 20 months between calving and sexual receptivity in females, heterosexual matings are quite rare, and auto-erotic behavior may have evolved in males to maintain sexual fitness. It is unlikely that necking giraffes think about this. They probably do it because it feels good.

Bats

Bats are enormously successful. Of the world's 4,000 or so mammal species, about one in four is a bat. There are two subgroups of bats – the Microchiroptera or 'microbats' and Megachiroptera or 'megabats.' Despite the obvious common feature of having fully-functional wings, the two groups are quite distinct in appearance, mainly because of their very different sensory systems – only micros locate using echoes, hence their huge ears. Bats' secretive, nocturnal lives make them challenging to study, so there have been few observations of their sexual behavior.

Gray-headed flying foxes are megabats. They typically roost in single-sex colonies, or 'camps.' Mutual grooming and caressing are common between pairs of individuals in both male and female camps. Pairs typically face one another, one's wings wrapped around the other while he or she licks, nibbles, rubs and/or grooms the fur of his/her partner. Males sometimes show erections during these interactions. Similar sessions happen between Livingstone's fruit bats, sometimes leading to homosexual mounting (but not penetration), one bat gripping and biting the other by the scruff of the neck from behind. Mouth-to-genital contact often occurs among gray-headed flying foxes. Both sexes may lick each other's genitals during mating. Males reportedly insert their tongue into a female's vagina for long periods.

Among the microbats, male common vampire bats lick and groom each other, often focusing particularly on the genitals. While licking another, a male may also rub his penis with a hind foot. Single-sex mounting has been reported in at least half a dozen echolocating bats.

Rodents and other mammals

Numbering over 1,200 species, the rodents are the most diverse mammalian order. Owing to their small size and a troubled historical relationship with humans, they are unfairly dismissed as 'lower mammals,' and little is known of their capacity for sexual pleasure apart from when they are used in the lab to study it, as we saw. Male dwarf cavies – small, guinea-pig-like rodents of South America – occasionally sit back on their haunches and make pelvic thrusts then lick and nuzzle their erect penises. During sex, male rats purr and females utter ultrasonic calls. Male rats trained to expect sexual contact with a female show an anticipatory response, exploring and moving about as they do in anticipation of being placed in a larger, more interesting cage. They show no such response when expecting a forced swim or being put in a small, barren cage. Male rats also prefer places where they have previously copulated.

Captive female elephants sometimes rub one another's clitoris with their trunks. Among the marsupials, male eastern gray kangaroos regularly thrust their erect penises repeatedly into their forepaws. In at least one captive study, female long-eared hedgehogs of Central Asia and the Middle East engaged in a variety of affectionate interactions, including sliding along each other's body, cuddling, intensive mouth-to-genital contact, and mounting (all presumably with deft skill, being such prickly characters). One female used her front paws to lift the hindparts of another before licking her genital area.

Birds

An enormous array of amorous activity has been documented in birds, most of it focused around courtship and mate attraction and not sex itself. But you are more likely to find what you are looking for, and as scientists increasingly view animals as sentient beings with minds, new discoveries come to light. Thick-billed murres – sleek, black seabirds that nest on cliff-faces – were recently spotted 'mating' with clumps of vegetation. And in 2001 a team of scientists from Sheffield published in the journal *Nature* what may be the first record of orgasm in birds. Whereas most birds, male and female, have an all-purpose genital opening called a cloaca, the male red-billed buffalo weaver boasts a penis-like appendage, which he massages against a female he is courting until, after about 15–20 animated minutes, he reaches what the investigators say appears to be an orgasm. 'It [sic] shuddered and its eyes glazed over,' one of the research team told a journalist. Weavers belong to the largest bird group of all – the passeriforms, which number several thousand species. The implication may be that male red-billed buffalo weaver birds are not unique among this group in their capacity for sexual pleasure.

Homosexual, autoerotic, and non-seasonal mating has been seen in more than 120 bird species in at least 37 diverse families. If all this activity reflects confusion on the birds' part, then that's a lot of baffled birds – sexual behavior costs time and energy. A simpler explanation is that birds are not unconscious slaves to evolutionary pressures, but have other motives – such as relieving sexual frustration and enjoying themselves.

Reptiles, amphibians and fishes

Virtually no consideration has been given to whether sexual activity might involve pleasure in the remaining major vertebrate groups. One reptile study deserves a mention. As reported by Martin Wikelski and Silke Baürle in *Proceedings of the Royal Society of London*, marine iguanas do something called 'non-ejaculatory "masturbation".' Small males freeze in copulatory poses to deliver sperm to receptive females quickly, before being displaced by a larger male. It may be the first time scientists have used the 'm' word in connection with a reptile. Evidence for sensory pleasure in the ectothermic vertebrates, including that summarized elsewhere in this book (see especially Chapters 3, 7 and 10), supports the idea that sex is rewarding for them, too.

Touch also figures in the courtship of certain reptiles. Herpetologists use the term 'titillation' to describe the way a male turtle vibrates his long foreclaws (feet turned outwards) against a female's face during courtship. Each titillation bout may last for minutes, and may be repeated for hours.

Many fish use touch in their courtship rituals, including mouthing, biting, butting, tail slapping and other bodily contacts. In cichlids and gobies, the prelude to spawning includes repeated rubbing of each fish's sensitive genital papillae and ventral surfaces over the nest, causing the papillae to become enlarged and erect. Genital contact is important for effective stimulation in the spawning of many fish and amphibians.

Conclusion

Currently, our knowledge of sexual behavior in animals, especially non-primates, is sparse. However, there is enough evidence at least to suggest that it is stimulating and stirring, not tedious and ho-hum. When we look closely we find that the animal kingdom is a sexy place, where carnality finds expression in many forms. Animals are neither priggish nor especially shy. We need only watch patiently, and try not to blush.

Chapter 7
TOUCH

Making contact with pleasure

> *... your life a sluice of sensation along your sides*
>
> D. H. Lawrence, *Fish*

Few who have petted or stroked or belly-rubbed a domestic dog need any convincing that the dog enjoyed it. Dogs seem especially to enjoy a scratch under the collar, where they can't reach. They close their eyes, go limp, hold their heads still. If standing, their tail may begin to droop as they relax, seemingly focused on the joy of sensation. Dogs solicit our touch, rolling onto their sides, forelegs cocked and hindlegs open, a clear demand for a belly-rub. When we stop they nuzzle our hands for more.

Darwin was one of the first scientists to recognize the importance and pleasure that animals attach to physical touch. He wrote in *The Expression of the Emotions in Man and Animals*:

> Dogs and cats manifestly take pleasure in rubbing against their masters and mistresses, and in being rubbed or patted by them. Many kinds of monkeys, as I am assured by the keepers in the Zoological Gardens, delight in fondling and being fondled by each other, and by persons to whom they are attached.

That touch is vital to the development of healthy human infants, and to the general health and well-being of adults and children, is widely acknowledged. It can bring comfort, reassurance and well-being in a way that seeing, hearing and smelling cannot, for unlike these, touch requires physical contact. It also has a psychological component. There is a world of difference between a tap on the shoulder from a loved one and the same tap by a hostile stranger. Touch can secure bonds, mend arguments and heal aches.

Touch is similarly important to many other animals. All vertebrates have a nervous system that receives tactile sensations from the body surface.

In birds, the follicle at the base of each feather is richly supplied with sensory nerve endings that appear to be stimulated by movement of the central vein, or rachis, of the feather. A flight feather also has several closely associated filoplumes, each of which is a long spike with a tassel of barbs at its tip, which appear to act as sensory detectors. Feathers are important tactile organs, providing continual feedback on the speed and direction of air movements. They also allow birds to sense the slightest touch.

In mammals hair and fur play a similar role, in concert with the skin, to form a surrounding sensory organ. So-called sinus hairs, such as the whiskers of cats and dogs, are found in almost all mammals (humans are a rare exception), each equipped with more than 2,000 sensory nerve endings. The skin of mammals is itself sensitive around the face, extremities and genitals.

The connecting touch

Touch is an important means of animal communication. It conveys messages akin to 'I trust you,' 'I accept you,' and 'I like you.' For example, members of banded mongoose packs 'wind themselves into writhing, squirming balls in the most enthusiastic of greetings' when they reunite after a lengthy separation, according to biologist and BBC television presenter Charlotte Uhlenbroek.

Close contact with others helps maintain group integrity in social animals. Meerkats – small, highly social members of the mongoose family – are touching much of the time, be it curled up in their burrows, or standing erect on their hindlegs surveying their surroundings in their characteristic fashion. By contrast, adult iguanas are largely solitary, but as babies they are gregarious, lying with their heads and tails resting on one another's backs. Gordon Burghardt, who studies them, recognizes that this behavior may have many evolutionary pluses (e.g. predator detection, temperature regulation), but he also notes that this doesn't exclude the likelihood that it 'may provide a feeling of security that, as with us, can be pleasurable and relaxing.'

Zebras have an ingenious means of standing in pairs which is not only tactile but effective in other ways, too. Two animals stand alongside each other facing in opposite directions, allowing each to rest his or her head on the other's neck, shoulder or back. In addition to the bonding value of this close contact, the position has obvious ergonomic benefits, for neither has to hold his or her own head up. It also allows the animals to reduce their available body surfaces to biting flies, which they obligingly swish away with their tails, while they cooperatively survey the full 360 degrees of the landscape for possible danger. Nuzzles, gentle nibbles and licks are an important component of zebra interactions. Vigilance and parasite control aid survival while tactile joy encourages and reinforces the useful behavior.

The skin of whales and dolphins is highly sensitive. In a study in which dolphins could request rewards by pressing plastic symbols on a keyboard with the tips of their beaks, some animals favored getting a rub to getting a fish. Baleen whales love to rub up against things. In some locations, whale-watchers have gained the trust of gray whales, who ride up against the sides of boats to have their bodies stroked and patted. Dolphins seem to enjoy simply moving closely in one another's company, quietly touching and exploring together. Orcas' skin is known to be exquisitely sensitive in some areas, being richly endowed with nerve endings around the eyes and face, the blowhole, and both male and female genitals. Face and blowhole sensitivity may help whales detect the water surface for timing of breathing.

Cats seem to like being grasped and massaged at the nape, perhaps because that's what their mothers did to them as kittens. Their dreamy, half-closed eyes and relaxed body recall the way dogs look when being rubbed beneath the collar. Plus, cats are not above nudging a stroker for more.

The joy of grooming

Fur protects mammals from cold, rain, dirt and parasites. To do so it must be kept in good condition. That's why mammals spend a lot of time grooming and cleaning their fur. Some aquatic creatures, like otters, beavers, muskrats and minks, anoint themselves with oils from glands in their skin. Birds waterproof their feathers in a similar way.

Since licking and grooming keep animals and their offspring clean, warm and buoyant, it's no surprise that they also find it comforting, as we like to wash and brush our hair, or have a scratch with a loofah. And just as the touch of another, at least to humans, can take pleasure to new levels, social grooming, or 'allogrooming,' is widespread in mammals, and many bird species practice its avian counterpart, allopreening.

Like many other mammal species that form social groups, Camargue horses regularly groom one another, which seems to have a pleasant, calming effect. When human researchers groomed them in experiments, the animals' heart rates dropped significantly – but only if the touch was directed at neck areas that are the preferred grooming sites in this species.

Cows often lick one another with their powerful sandpapery tongues. Rosamund Young, a lifelong organic cattle farmer in Britain and author of *The Secret Life of Cows*, watches her cows groom daily. Most common is mothers licking their calves, but Young sees plenty of other grooming involving unrelated animals. Cows also enjoy being groomed by a trusted human. Young carries a brush for soothing disturbed cows who, for instance, have a foreign object lodged in a foot; grooming relaxes the animals enough for Young to remove the offending object. On one occasion, a cow named July so abandoned herself to the pleasure of Young's grooming that she fell asleep.

Chimps and other primates spend on average about 20% of their waking hours grooming one another. It has been shown that when rhesus monkeys groom each other, endorphins (chemicals associated with pain relief and reward) are released into their bloodstream. Some species form grooming lines of several individuals each, removing parasites, salt crystals and other material from the fur of the animal in front.

I keep a specially designed brush for my cats. Their pleasure in being coddled with it is unmistakable. I signal my intention to brush them by thumping the side of the brush on the carpet. Our rather portly marmalade male Walter does not walk to be brushed – he runs, his fat belly flopping side to side as he approaches. Camille, a graceful tabby, also can't get his brushing soon enough, and scampers to the signal. For all their aloof reputation, cats can show their pleasure well, and these two males hold their tails erect, purring loudly as the bristles rake across their backs and sides. Cats in this blissful state like to rub their muzzles forcefully against something, and my socks and knee are a favorite target. Cat pals often groom each other, using their raspy tongues to comb the other's fur. Sometimes these allogrooming sessions evolve into a wrestling match when the receiver decides he has had enough.

Preening is the bird equivalent of mammals grooming. Most birds spend a significant portion of their time combing feathers through their bills and applying water-resistant oils from a gland near their tails. Interpretations of allopreening have mainly considered plumage maintenance, parasite control and sustaining pair bonds critical to ensure the successful rearing of chicks. It may be tacitly assumed that birds find pleasure in it, but this is not generally spoken of.

The benefits of allopreening predict that it is pleasurable for the participants. The mated pairs of many bird species nibble and fuss over their partner's head and neck feathers. Parrots preen their mates for hours between naps, according to ornithologist Joanna Burger. Courting gulls and gannets share gentle, tender caresses. Preened birds appear relaxed and appreciative. During a recent field study of trumpeters – highly social birds which look rather like coots and forage for fruits and snakes along the floor of humid South American forests – biologist Joseph Tobias observed their allopreening behavior on many occasions, describing it as follows:

> A chin or nape is nibbled in a seemingly affectionate manner. The recipient falls into a heavy-lidded trance, which looks very much like rapture.

Figure 7.1 Grooming releases reward compounds in a macaque's brain.

Birds kept as human companions often preen their guardians. In *The Human Nature of Birds*, psychologist Theodore Barber reports an orphaned jackdaw raised by hand in England expressing affection by rubbing his bill back and forth around his human companion's ear. A parakeet observed over the course of years by the ornithologist Cheryl Wilson would 'ask' Wilson to groom his feathers in the same way that he did his parakeet partner Blondie – by slightly raising his feathers and tilting his head in a particular way.

Even birds of prey, who may give the outward impression of being all business and no pleasure, appreciate the tender touch. While rehabilitating a tawny owl hit by a car, Mike Tomkies describes how 'he loved being tickled under the chin, stretching himself to his full height of just over a foot, his eyes half-closed sleepily.' One could be forgiven for thinking Tomkies was coddling a cat.

Interspecies touch

The pleasures of inter-individual touch are not limited to members of the same species. The most familiar form of interspecies touch is our connection with the companion animals with whom we share our homes. Indeed, that we refer to them as 'pets' denotes the historical importance of touch to these relationships. We love the feel of a cat's fur and the solidity of a larger dog's torso. And they seem to relish the tactile affections we lavish upon them.

Interspecies touch shows up in some unexpected places. In the Galapagos, finches groom giant tortoises. The birds land in front of the large reptiles and hop up and down to signal their desire to forage. If a tortoise is amenable, she extends her neck to its full length and stands stock still with stiff legs, exposing the many nooks and crannies where mites or other parasites might be lurking. Both bird and reptile benefit, and they behave in a manner to suggest that each recognizes as much. It is reasonable, therefore, to infer that both find pleasure in it.

Observations in captivity corroborate the tortoise's liking for the gentle touch of another. At the Philadelphia zoo, one elderly Aldabran tortoise will plod over to his long-time attendant and rise up on all four legs for a neck scratch. Only the one attendant is sought out for this attention, regardless of his clothing. Other Aldabrans there also solicit neck scratches, and one will even nip at the attendant's leg if the advances are ignored. Bonnie Bowers and Gordon Burghardt, the ethologists from the University of Tennessee who reported these interactions, detail similar behavior in their captive iguanas, who cock their heads to be scratched and rubbed, and close their eyes while extending their front legs and dewlaps.

Wildlife cinematographers Mark Deeble and Victoria Stone constructed an ingenious underwater viewing window through which they watched and photographed hippopotamuses in an African fresh-water spring. They soon discovered that here, too, a pact had been established between the hippos and fish who came to clean them. The hippos were 'far from passive participants' in the cleaning services of several species of fish:

> We saw them [hippos] deliberately splay their toes and spread their legs to provide easy access or to solicit cleanings. They would even visit 'cleaning stations' where

> fish congregate – much like pampered clients going for a massage or manicure at a spa.

Once Deeble and Stone identified the regular hippo cleaning stations, they found that several fish species were partitioning a hippo's body among them. Barbels used their pointy snouts to get between the toes and under the tail, cichlids groomed the bristles, while tiny *Garra* cleaned wounds and gashes on the skin. *Labeos* acted as toothbrushes, polishing the hippos' teeth in response to their jaw-dropping. The hippos were so relaxed during these piscine pedicures that they would sometimes fall asleep.

Central American mice live with large rove beetles. The beetles ride around on the mice's fur during the day and hunt fleas in their nests by night. Forty species of rove beetle provide this service for an equally diverse mouse clientele. Why do the mice tolerate these hitchhikers? Do they realize that the beetles provide parasite control? Or do they perhaps enjoy the tickle of the beetles' movements? Evolutionary theory would suggest that because superior parasite control can translate into enhanced reproductive success, natural selection favors mice who don't remove rove beetles. This explanation ignores the role of conscious experience. That the mice do not groom out their coleopteran companions suggests that they at least tolerate them, and perhaps welcome them.

Small moths dwell on the fur of sloths, awaiting the opportunity to lay eggs in the sloth's dung when they descend to the ground to defecate. The relationship is thought to benefit only the moth. I have seen film of these moths crawling across a sloth's face; the mammal closes his/her eyes but appears otherwise unperturbed, if not contented, to have these guests. If the moths were a nuisance, why would the sloth be so tolerant? Do sloths ever scratch to try to remove them? Is the mammal simply incapable of repelling the insects, or might there be some pleasure to be had from the gentle tickles made by the scuttling moths' feet? It would certainly be beneficial to the moths if so.

The wildlife photographer Hugo Van Lawick once watched several agamid lizards clamber on the backs of a group of resting lions, where they caught many flies. The lions twitched from time to time, but otherwise appeared to take no notice of the lizards. Van Lawick suspected that the lions were in fact glad to have these small predators relieving them of the bothersome insects.

Many mammals (impalas, giraffe, hippopotamus, capybara, etc.) are accompanied by birds who ride on their backs and feed on stirred up insects, ticks, earwax, dandruff and even sips of blood. Despite the occasional parasitic tipple, the mammalian hosts are very tolerant, and appear to enjoy the birds' attentions. Perhaps this tolerance stems from the birds' parasite-removal service, or from their aid in predator vigilance, or perhaps the hosts simply enjoy the touch. Or all three.

Dr James Morgan, a zoologist, sheep farmer, and self-proclaimed 'armchair ethologist,' shared with me his subtle but intriguing observations on the interactions of ewes and brown-headed cowbirds:

> When a female cowbird landed on a ewe, the ewe immediately quit grazing and stood very still. These ewes remained still, even though they were quite ravenous (they were on a restricted pre-breeding diet, and would practically stampede me when I opened the pasture gates). As I went about my chores and came too close, the cowbird would fly to another ewe. Immediately, that ewe would stop grazing, stand very still and relaxed, as if mesmerized. If I was to assign an emotion to the ewe, I would say she was acting incredibly contented when the cowbird was on her back.

Sometimes the connections between animals of different species take us completely by surprise. An apparent friendship developed between a leopard and a cow in the village of

Wghodia Taluka, India. According to wildlife warden Ro-hit Vyas, who has visited the village several times with other animal enthusiasts, the leopard had been visiting the cow regularly for months when their story appeared in *The Times of India*. When they meet, 'the fearless cow licks the leopard on its head and neck.'

It is taking a long time, but the importance to animals of physical contact is gradually being recognized in industries inherently exploitative of animals. In Germany, government officials involved in the enactment of animal welfare regulations are encouraging pig farmers to give their pigs 20 seconds of human contact each day, on the basis that a little tender loving care is good for health and welfare. This is hardly lavishing on the affection, but it is significant for what it symbolizes – the notion that an animal destined for the slaughterhouse still deserves respect and compassion.

Scratching beneath the surface

Relieving an itch by scratching is one of life's private pleasures. We find bliss in the feel of a loofah sponge, a back scratcher, a dry brush or the gentle raking of our own or a close friend's nails across the back or arms. Humans are not the only creatures that scratch themselves; the behavior is widespread in the animal kingdom. Female eastern gray kangaroos form pair bonds, the partners of which frequently groom one another, affectionately licking, nibbling and raking the fur on the other's head and neck with the paws. Practically all terrestrial mammals, but most familiarly cats and dogs, will use a hindpaw to relieve an itch, and I have watched in amusement as a particularly loose collar did a full circuit around a scratching dog's neck.

Scratching may have adaptive value, such as for the removal of parasites or sloughing skin. For the urgent relief of an irresistible itch it could even be said to preserve one's sanity. Perhaps itchiness evolved to promote scratching, given the various benefits it may provide. Scratching undoubtedly removes parasites,

Figure 7.2 Scratching gives a sea lion delicious relief from an itch.

but its unmistakable resemblance to our own scratching behavior strongly suggests that it is pleasurable.

Elephants scratch themselves with sticks grasped in their trunks. Many animals use the sandpapery surface of rocks and termite mounds, or the rough bark of a tree, to rub and scratch. I've seen BBC documentary footage of a capybara (an enormous South American rodent) having a good rolling scratch on the ground, much as dogs do.

Mirthful rats

Because of the negative folklore surrounding them, rats are less understood when it comes to the pleasure of touch. But like cats and dogs, these social animals will nip gently to solicit more tickling. Noted neuroscientist Jaak Panksepp and his colleagues at Washington State University have for many years been studying rat feelings (or 'affective states,' to use the jargon). In 2001 they published in the journal *Physiology & Behavior* the results of a series of experiments on rats' responses to being tickled. Tickling consisted of vigorous playful whole-body stimulation with small, rapid movements of the fingers of one hand. Rats were gently pinned on their backs and the tickles directed at their undersides, in the manner of playful tickling in humans.

In the first experiment, rats were assigned to one of either 'tickling' or 'petting' groups, then individually placed in a chamber. There they received either tickling or petting (repeated gentle touching of the animal's back) for 15 seconds at 15 second intervals over two minutes on each of five consecutive days. During these sessions, the rats' calls were recorded to compare the number of 50 kHz calls, which rats make during pleasurable situations. Tickled rats made seven times more 50 kHz calls than petted rats. The difference also increased over the five-day period, suggesting that the rats' enthusiasm for being tickled grew.

If the rats were truly enjoying being tickled, then we might find that they hurry to a place where they expected to be tickled, just as my cats will run to be brushed. And so they did. The investigators measured the time it took rats to approach a human hand from the opposite end of the tickling/petting chamber, a distance of about 20 inches. Rats accustomed to being tickled ran to the hand four times as quickly as petted rats. It would appear that while petting feels good to rats, being tickled is a blast!

In a further experiment, rats were trained for nine days to press a bar to receive tickling. Following these nine days the rats received no tickling when the bar was pressed; this 'extinction testing' was done to reduce the possibility that the rats might bar-press merely out of habit during the experiment that followed. Next, each rat was presented with two bars, only one of which elicited tickling when pressed. You are by now probably expecting that the rats pressed the tickle bar much more than the other bar. Well, they did. They pressed the tickle bar repeatedly, and the other bar almost never.

This study is a great example of scientists' potential to reveal aspects of animal pleasure.

Feeling like a fish

Humans tend to dismiss fish as lacking intelligence and feelings. This attitude is a product of unfamiliarity and expedience, given that our closest contact with fishes is usually either at the end of a line or a fork. Those who take more than a gustatory or recreational interest must usually depend on documentaries and descriptions to catch glimpses of fishes' private, natural lives. Our relative ignorance of marine life is further illustrated by the fact that approximately three new fish species are discovered each week. We have much to learn about them.

Fish are sensitive to touch. They are equipped with sensory organs; the lateral lines of bony fish and the organs of Lorenzini (jelly-filled pores in the skin that detect minute electrical fields) in sharks and rays, for instance, detect pressure changes and flow patterns in the water.

There is at least one example of a fish receiving the ministrations of a bird's beak. Ocean sunfish come to the surface and float on their sides, permitting gulls to remove fish-lice from their flanks with their beaks. Phalaropes have been seen grooming whales in a similar way. Fish who have grown accustomed to the presence of divers sometimes permit the humans to stroke them. These animals don't seem to gain anything from these interactions in terms of food or other basic survival benefits, which increases the likelihood that they do it because it feels good.

The widespread phenomenon of fish cleaning stations offers compelling support for the experience of tactile pleasure in fish. Cleaner fish of a variety of species nibble loose skin, fungal growths and fish lice from other fish. Cleaners also pluck at wounds, which appears to relieve infection and speed healing. It's a definitive mutualism: cleaners get food (delivered buffet-style by clients who line up patiently to await their turn) and clients get a body-spruce-up service. Different species are picked over in a highly specific manner by cleaner fish, who advertise their services with brightly colored uniforms, and perform bobbing/fussing movements to signal their willingness. Clients also may signal their readiness, for instance by orienting themselves vertically in the water, and opening their mouths and gills at appropriate times to allow access to the cleaner fishes. As the old saying goes, 'the sign brings customers,' and these behaviors may help convey the benign intent of the customer towards the usually much smaller cleaner, just as play-signals help ensure that a play-fight isn't confused with a serious fight (see Chapter 4).

Features of the cleaner–client relationship suggest positive feelings are involved. Invitation postures indicate that cleaners may be anticipating the attentions of clients, and vice versa. Some fish change color while being 'serviced' by cleaners, a behavior that has been associated with changes in emotions (e.g. stress, arousal); that clients seek out cleaners rather than avoid them is consonant with a pleasurable, and not a negative, stress response. Recent study of one of the principal cleaner species, the cleaner wrasse, supports the notion that tactile stimulation is an important motivator for the interaction. One-hundred twelve hours of surveillance of 12 different cleaners revealed that cleaners appear able to alter client decisions over how long to stay for an inspection, and to stop clients from fleeing or responding aggressively to a bite that made them jolt.

There is also evidence that cleaners and clients recognize each other, and that they return to their favored business partner, much as we return to a favorite barber or hairdresser. Then there is the patience and dedication shown by the participants. A cleaner wrasse may have more than 2,300 interactions per day with various individuals of various species of clients, and there still may be a queue!

Like so many arrangements built on trust, this one is open to exploitation. Some species of fish, such as the sabertooth blenny, mimic cleaner-fish in both appearance and posture; then, when the customer is least expecting it, the little Sweeney Todd bites off a chunk of fin or flesh and darts away.

Getting comfortable

Many animal pleasures can be traced to facets of their individual life histories. The kneading behavior of adult cats is probably a remnant of infancy, when kittens knead their mother's teats to express milk. Kneading cats often have their mouths close to their paws, much as nursing kittens do. The animal alternately pedals the forepaws against a favorite blanket or a human companion's belly (hopefully well insulated), eyes closed, purring loudly. Kneading cats exude contented absorption.

Chickens and other birds often become calm once cradled in someone's arms even though they may have struggled to avoid capture. This behavior could derive from the comfort of being incubated as a chick. If left alone, baby chickens will cry pitifully for hours, but if held gently in one's hands, they show a comfort response, rapidly closing their eyes and falling asleep.

A friend reported to me that one of her cats sucks his tail during quiet moments of comfort. Trunk-sucking is common in young elephants, and is interpreted as a comfort behavior akin to thumb-sucking in human children. Elephants also sometimes greet one another by putting tips of trunks in each other's mouths. Might a tail or a trunk be a nipple substitute, helping to stimulate nostalgic feelings of nursing in infancy? In *The Naked Ape*, Desmond Morris theorized that the physical comfort of cigarette and pipe smoking may derive from the infantile pleasure of nursing at the breast. My point is that the animals' behaviors suggest a calming effect, and that they might be pleasurable, as they can be for us.

The pleasure of comfort is in avoiding unpleasant or uncomfortable things, such as temperatures that are too cold or hot. On a windy October day in New Jersey I watched a flock of about 70 sanderlings resting in a tight huddle on a sandy beach just beyond the waves' reach. Each bird's posture was the same: standing on one leg with head tucked across the back, bill inserted under the wing border. Every few moments, individuals would change positions, perhaps to get a safer position away from the periphery of the huddle, or to get more rays from the soon setting sun. When I saw the first bird do this, she hopped, and I thought she must be missing a leg. While roosting on one leg is common for these and other shorebirds, when they run along the beach skillfully avoiding the surf between probing assaults on sand-bound invertebrates, they invariably use both legs. To my surprise, however, I noticed that whenever a bird changed positions in this huddle, he/she almost invariably hopped. Unless this was a support group for amputees, they were all keeping their other leg tucked away. But why hop when you can use two legs?

An evolutionary explanation for this behavior might run something like this: the energy expended by deploying the second, tucked-in leg, might exceed the energy expended by keeping it tucked in. Deploying leg #2 may lose body heat, and even though hopping is less efficient than walking/trotting, the distance to the new resting spot in the flock is short enough to more than negate the use of the second leg. A more sensual pleasure-acknowledging explanation might go like this: a sanderling who decides to shift positions in the flock prefers not to deploy his or her other leg for reasons of comfort. That foot is warm and dry and it would be unpleasant to get it wet. These two explanations – evolutionary and experiential – are perfectly compatible.

Hugo Van Lawick describes the rather comical sight of a group of East African wild dogs lying in a heap and trying to keep warm while a cold wind blew across the plains. Intermittently, dogs on the windward side would get up and move to a leeward spot. Over the course of several hours, the huddle had 'migrated' to a completely new location. Newborn mouse litters form similar 'dynamic huddles,' in which pups continually burrow into the center of the cluster, conserving heat and gaining comfort.

Dog and cat lovers will know that these animals seek warmth like a wave seeks the shore. It is no coincidence that they spend more time taking up leg-room on the bed during winter than during summer, or that a sweater straight out of the dryer is a cat magnet. Little Bit, one of several American black bears befriended by Jack and Patti Becklund in northern Minnesota during the 1990s, decided that a pink bathrobe made a comfy den bed. After Patti noticed it missing from the clothesline, she found it in the woods with a tell-tale whorl of black fur in the middle. Later, after washing and re-hanging it, the bathrobe went missing again, and this time the culprit was still using it when Patti came on the scene. The robe was Patti's favorite, thick and plush. Apparently, Little Bit found it equally to her liking. Perhaps the robe's allure was more related to its familiarity or its association with a trusted friend than with its softness, but either way it seems to have lent comfort to the bear.

Elephants sleep for relatively short periods at a time, which prevents their inner organs being damaged by their huge mass. As a further protection, they make pillows, gathering together grasses, twigs and leaves to cushion their heads and flanks. They do it for comfort; it's another case of what provides survival benefit also feels good.

In John Paling's 1979 book *Squirrel on my Shoulder*, the caption next to a photograph of a gray squirrel splayed out on a branch in the sun reads: 'A squirrel sunning itself on a tree branch – this is thought to produce vitamins in its fur.' This caption implies that the animal suns only because if s/he didn't s/he would become vitamin-deficient. I would hazard that the squirrel suns because soaking up the warmth feels good. When you hunker close to a campfire on a cold night, you may be keeping your temperature at a disease-resisting level and saving enough energy to make a difference between life and death if you became stranded, for instance. But you don't get close to the fire to survive; you get close because the warmth feels nice. The same goes for the wet cormorant drying her wings in the sun. Dry wings aid flight, and the pleasure of the sun's warmth rewards and motivates the valuable behavior.

Sunning is another source of comfort. Birds fan their wings out to get the most of the sun's rays. I have watched a blue jay searching around on a bed of dried leaves on a steep south-facing slope. Having found a suitable spot, she then turned facing up the slope, pressed her breast to the leaves, and spread out her wings and tail feathers to face the direct sunlight. She remained there for about one minute before rising to pursue an insect. I recently startled a starling while cycling across a playing field; she was spread so flat against the grass that she didn't see me – nor I her – until the last minute.

The opposite of not enough sun is too much. Animals often seek ways to cool off. Shade is one solution. I once photographed a herd of about 20 impala in the midday sun of Kruger Park, South Africa, crowded beneath the shade of a single tree. The website of French photographers Nicole and André Brunsperger includes an extraordinary photo of nine lions resting in a tightly packed line beneath the shade of a small airplane wing.

Water also provides relief from heat. Australian Reg Clark, a man whose zest for birds is matched only by his zeal for gardening, described to me the afternoon he was watering his petunias in Sydney when he noticed a movement in a nearby bush. It was a brown cuckoodove. Man and cuckoodove regarded each other with mutual caution for a minute or so. Then it occurred to Reg that the bird might enjoy a gentle shower on this 38° C day. Adjusting his hose to a fine spray, Reg slowly directed the water towards, then onto, the shy creature. The cuckoodove's immediate reaction was to close his eyes and stretch forward into the moisture, fluffing out the feathers on his breast and abdomen. He held this pose for some minutes. He then settled back on the branch and, to Reg's surprise, leaned over sideways to a 45° angle and extended his left wing vertically, allowing Reg to thoroughly spray his 'wingpit.' This went on for two or three more minutes. Then, maybe realizing that one can't have too much of a good thing, the bird leaned over in the opposite direction and raised his right wing, permitting Reg to soak the other side. All the while this was happening – about five minutes – the brown cuckoodove appeared to have his eyes closed and, to Reg, looked in a state of sheer bliss. Finally, the bird turned around to have his back watered with all feathers raised, then shook violently several times and flew off.

I once read in an American backyard birdwatcher newsletter a similar description of two mourning doves sitting in the path of a sprinkler. Every time the sprinkler would come by, in unison, the birds lifted a wing in the air and rolled to the side, allowing the water to reach their flanks. They did this for about ten minutes, sometimes rolling to one side, sometimes the other.

Joanna Burger describes the 'glee' she's observed in a small mixed flock of scarlet macaws, chestnut-fronted macaws, and blue-headed parrots during the first rain in about three months; the birds screeched loudly, stretched their wings, and held their opened bills skyward drinking in the rain. At times they swung upside down.

A recent account from two ornithologists in the bird journal *The Wilson Bulletin* describes a flock of about 15 nighthawks flying in a torrential midday summer downpour. Normally this species hunts for insects on the wing at dusk and dawn. They were not foraging, and they glided in place, facing into the wind which gusted up to 37 km/hour. Their body plumage was ruffled, their tail feathers spread slightly, and their wing beats slower than normal. They flew in the same spot for about 10 minutes before heading west when the rain slackened. The authors note that many birds bathe in rain or wet foliage, and that aerial rain bathing is probably important (and, I would add, refreshing) for species poorly built for bathing in standing water.

Conclusion

Touch is the most physical of the senses. Because life on Earth inevitably requires being in contact with something (ground water, or air), evolution has equipped animals with the ability to sense and respond to that contact. The animals many of us are most familiar with – cats and dogs – show remarkable sensitivity to touch. The slightest touch on the hairs of a napping cat's ears or a sleeping dog's eyelids produces a reflexive flinch. These same animals luxuriate when stroked or rubbed in the right places. Watch how a fish darts away when it is touched on the flank, and you'll know just how sensitive they are to touch. We associate such responses with survival and escaping danger. Yet touch also soothes, excites and comforts. It rewards relieving an itch, removing a parasite or maintaining well-oiled fur and feathers. It helps secure and maintain social relationships, and it facilitates courting and mating. A world devoid of touch would be as much a loss for the other animals as it would for us. If you doubt it, go rub a dog's belly.

Chapter 8
LOVE

The ripening warmth of intimacy

Everyone who has domesticated some shy creature can testify to the wealth of character it came to display in the ripening warmth of intimacy.

Colonel E. B. Hamley, *Our Poor Relations*, 1872

On a trip to Edinburgh in 2003, I visited the statue and fountain of Greyfriars Bobby. Bobby, a Skye terrier, was the close companion of John Gray, who died in 1858 and was buried near the site of the present-day statue. Though dogs were not at that time officially permitted on the cemetery where Gray was interred, Bobby refused to be discouraged by the cemetery keeper's efforts to keep him out. On the third day after Gray's burial, as Bobby lay at the mound of earth during a cold rain, the keeper took pity and brought him some food. Thereafter Bobby had free access to the cemetery, and for the next 14 years the little dog never spent a night away from his master's tomb. The fountain and statue commemorating his devotion was erected near the gravesite in 1873, and Bobby has since gained legendary fame as a symbol of love and devotion.

Bobby's behavior was exceptional but not unique. Dogs are highly social and fiercely loyal, and examples of dog-loves-man scenarios abound. But are Bobby and his ilk really experiencing some sort of love towards their human companions? The little Scottish dog's behavior suggests a strong and enduring attachment, but love is a difficult enough concept to pin down in humans, let alone other animals. Bobby's behavior might as soon be described as loyalty, devotion, dependence or some other feeling. Then again, all these emotions are variations on the same theme, and as we saw in Chapter 1, evidence from studies of animal and human brains, and their behavior, suggests common origins across many animal groups.

Adaptive love

Should we expect animals to display feelings of love? Is love adaptive? It should be for some species. We may expect love to evolve in animals where a close, lasting pairing is beneficial, such as where the successful rearing of young requires both parents to cooperate. Ninety per cent of bird species are believed monogamous, and about half mate for life (not bad when one considers that close to half of British and American marriages end in divorce). Parrots, geese and penguins are among the more notable examples.

Many mammals also form monogamous pairings for a breeding season or longer. There is evidence that could be interpreted as love among whales, dolphins, hyenas, horses, foxes and mongooses.

We may also expect to find love in strongly social species, where being part of a tight-knit group helps group members to find food, detect and deter predators, or keep warm. Prairie dogs, baboons, meerkats and rabbits spring to mind. Close attachments between individuals may also help to achieve or maintain higher social status. Both male and female chimpanzees, for instance, form long-term alliances with members of the same sex that benefit their social status in a group.

And we should not exclude the likelihood that animals develop close emotional attachments to others in their group regardless of any ultimate, evolutionary benefit. They may just like each other, as close friends do. According to Jane Goodall, love in chimpanzees is expressed as sympathy, tenderness, joy, understanding, and other emotions shared between close individuals. Love's evolutionary cousins in these apes include friendship, greetings, the pleasure of relaxed contact, food sharing, and helping and soothing those in distress.

A barn swallow was recently reported to remain with his fallen mate, who lay apparently dead after colliding with a window. Five minutes later the stricken female roused, recovered and the two flew off together. This touching scene stirs the hearts of sentimental human onlookers. A biologist might caution that the male was simply trying to secure his reproductive investment. Perhaps. The same could be said of a parent who swims out to rescue a drowning child, but hard-line evolutionary interpretations would hardly convey the whole picture. Feelings are involved, and it is these feelings that motivate and reinforce the behavior. Evolutionary bases for the existence of long-term animal partnerships support rather than undermine the importance of emotions in helping to seal the partnership. If it benefits survival that two birds stay together, then sometimes mutual feelings of love and attachment are also adaptive.

Love is of course not just about the good times. How animals respond to the loss of a friend, as opposed to a mate, is another measure of their emotional connection. A video filmed by a respected whale researcher suggests a depth of love or affection shared by two humpback whales. One of the whales had recently died, and the other repeatedly swam up beneath him, gently nudging him to the surface as mother whales do to help their calves take their first breath. The living whale floated motionless beneath the other before rising up and embracing him with his huge pectoral fins. The embrace lasted five hours. This is a single anecdote, but a telling one. Marine biologist Bernd Wursig witnessed a pair of right whales off the coast of Argentina, each stroking the other with their flippers, rolling in what looked like an embrace, then remaining touching as they swam away slowly, diving and surfacing in unison.

Loss may bring profound grief. Cats' and dogs' demeanor can change markedly when a feline or canine friend dies. Geese can live 50 or more years and they mate for life. Konrad Lorenz was among those who had no doubt that geese mourn the loss of a partner. He described the overall drooping appearance, the hanging head, and the eyes sunk deep into the sockets of bereaved greylag geese.

Recently as I sat one evening reading on the banks of the Ouse River in York, a family of moorhens swam upstream nearby. Two tiny fluffy black chicks were paddling hard to keep up with their parents. Then, as they reached the far side of the river a little upstream, the parent birds suddenly produced a series of mournful sounding cries the likes of which I'd never before heard from moorhens. They called for about 20 seconds before making their way to a nearby floating perch, at which point I noticed that they now only had one chick in tow. Something had suddenly befallen the other one and the parents' strange calls seemed to signal their reaction to the event. It is of course as anthropomorphic to interpret their reaction as an expression of their loss as it is to ascribe a mournful quality to their cries. But is it so far-fetched to think that these birds have an emotional attachment to their hard-won chicks? The skeptic may respond with accounts of cruelty to chicks by coots, which are close relatives of moorhens, and which are known to shun and sometimes kill their own chicks in times of hardship. But it isn't clear to me why one sort of behavior precludes the possibility of another. Infanticide and mutilation of infants occurs in human societies, yet we do not question our species' love for our children.

The mated pair

Loving behaviors help to maintain loving bonds. We send cards or flowers to remind loved ones of our feelings towards them. Other species also demonstrate their devotion to each other. Many species of birds, once bonded as a pair, will perch close together, preen each other, share food and rarely leave each other's side. Like their jackdaw cousins, ravens too pair for life. Bernd Heinrich believes ravens experience love:

> I suspect they fall in love like we do, simply because some kind of internal reward is required to maintain a long-term pair bond.

Like many other birds, ravens produce more young when both parents are at the nest, so evolution will have favored rewarding, pleasurable pair-cementing interactions. We can fairly safely assume that ravens are not consciously aware that producing more offspring will secure the future of their genes. It is the pleasure – the rewards of love – that provides the palpable motivation for pair-bonding.

Many birds perform 'bill-fondling,' in which two individuals gently nuzzle each other's bills. It looks like a kiss, and probably functions similarly – to feel pleasure and to strengthen a pair-bond. It is practiced by many birds, including parrots, babblers, waxbills and mynas. I have seen it in members of the pigeon and finch families. I observed a male American goldfinch land on a ledge just outside my office window. He was followed about 20 seconds later by a female. During the next two minutes, the female made two gapes towards the male, opening her bill half-way for about a second each time. Then she hopped closer to him and the pair engaged in two or three bouts of bill-fondling, each leaning towards the other, bringing their bill tips together and making nibbling motions. They oriented their bills at 90 degree angles, as kissing humans typically do.

A year later, as I walked across one of the medieval bridges that span the River Ouse in northern England, I noticed a pair of rock doves atop one of the bridge's stone columns. Rock doves are as common as rooftops in York, but what caught my eye with this pair was that they were kissing. So intense was their facial embrace that as I walked past them about 15 feet away, I began to worry that perhaps they had become stuck, in the manner of two dogs locked in copulation. A few seconds later they broke off. They had two more 'necking' sessions, during which I noticed rhythmic movements on the neck feathers of the two birds. One was probably sharing the contents of his or her crop with the other, a behavior called 'courtship feeding' by ethologists. The smaller bird then crouched down with her head and body low to the ground, the male mounted her back and they copulated before flying away.

A pair of voles nested under a feeder in Scotland into which wildlife writer Mike Tomkies put food for the birds and mammals he watched and photographed through his window. He describes the mates coming up to feed after the female had given birth:

> They rubbed muzzles together gently before he passed [corn]flakes to her with his front paws.

Oxytocin, a hormone implicated in social bonding in humans, has also been shown to be important in the formation of social friendships and allegiances in species exhibiting adult pair-bonding, such as prairie voles, for mother–infant bonding in sheep, solidification of social memories in rats, relief of separation distress in birds, and maternal urges in many species.

Love across the divide

Too often the connection between parent and offspring – which may be as nurturing as that between a human parent and her or his child – is seen in the animal only in the context of survival and reproduction. Yet, as Masson and McCarthy lament, herein lies a double standard:

> In considering if it is possible to know whether a mother [non-human] ape loves her baby, it is worth asking if it is possible to know whether the people down the street love their baby. They feed it and care for it. They tickle it and play with it. They defend it with all their might. But all that is not considered proof in the case of the ape.

Nature is replete with signs of parental love. An orca mother grants a watchful latitude for her calf to explore and play. A chimpanzee infant cradles and grooms a surrogate log in a transference of her mother's love. A peregrine falcon strips small pieces of meat from a kill and delicately proffers them for her hungry chick to snatch from her bill. Can we doubt the love a cow feels for her calf when she may bellow despondently for days after her baby is taken away (so that all – not some – of her milk can be taken for human consumption)?

In many species, fathers care for young. Paternal care is known in about 10% of mammals, and in many birds and fish. Conventionally, scientists attribute such behavior to increased offspring survival. Paternal care evolved in species for which the demands of raising the offspring successfully are particularly high, where only one parent may fail to cope. If fatherly dedication and bi-parental care are important to successful rearing of young, then natural selection should favor doting fathers.

When humans raise children born of other parents, as happens in adoption, say, they invest in someone with no genetic relation. Such is the power of the emotional experience of raising a child that human parents yearn for it regardless of evolutionary benefit.

We are not alone. Animals also adopt. Adoption, foster-parenting, 'alloparenting' and even kidnapping occur in many mammals, including various apes, monkeys, dolphins and whales, seals, rodents, and carnivores. Cooperative breeding, in which one or more individuals forego reproducing and instead help to raise the young of related and unrelated parents, is especially widespread in birds, but also occurs in many mammals, and at least two types of fish. There are solid evolutionary theories to account for an animal's decision to become a cooperative breeder, such as making the best of a situation where nesting, mating or food resources are scarce, and improving one's future

Figure 8.1 A ring-tailed lemur lavishes parental love.

chances of inheriting a good breeding territory or mate. None of these ideas need exclude the role of emotions; the evolution of cooperative breeding should favor feelings of attachment and affection towards the young under one's care.

Courtship

Walt Disney's 1942 film *Bambi* depicts several animals becoming 'twitterpated' with each other. While anthropomorphic to a fault, these brilliantly rendered scenes project onto the animals the sense of euphoria we feel during our own courtship. Elaborate courtship rituals are widespread in animals, including all of the vertebrate classes, and many invertebrates.

Ethologists have offered many theories for the evolution of courtship behaviors. Prolonged pre-mating interactions may help confirm the evolutionary fitness and compatibility of potential mates. One sees some parallels here with human courtship, which also functions to assess such qualities as fidelity, reliability and compatibility. Quite a few male birds, including northern cardinals and blue jays, feed tender morsels of food to their ladies. Sometimes the females will assume a fledgling begging posture while being fed. Other birds, such as the roadrunner, may bring food to the female just before mating. Not surprisingly, species of birds that feed their girlfriends usually are the guys who stick around to help feed and raise the young chicks. These are important considerations for choosy females.

Just as consorting with a new partner is usually a heady, exciting experience for us, so it may be for other species. When two courting sarus cranes make high flapping leaps several feet into the air, it could be that their hearts also leap. That is an anthropomorphic hunch, but the burden of proof could lie more with those who would deny such feelings in these animals than it currently does. Notwithstanding that it *looks* like the birds are enraptured by each other, the idea that evolution should deal feelings of euphoria only to humans is unlikely.

Companionship

In August 2002 I watched three adult female common mergansers patrolling the edgewaters of a small bay in the Adirondacks of New York State. In the same manner I had observed 22 juvenile red-breasted mergansers (and one chaperoning adult female) on Lake Muskoka in southern Ontario in 1999. In characteristic merganser fashion, they stayed close to the shore, swimming generally in one direction, their faces usually pointed beneath the water surface looking for fish.

One may ask why these three should stay together when they are competing for limited food resources (I had snorkeled these same edgewaters earlier that day and can vouch for the relative scarcity of fish there). There are benefits to such filial behavior, such as predator detection and avoidance, and cooperative foraging. Another way of looking at it is that these were three buddies, or perhaps siblings who shared a loving family bond dating back to their infancy.

Nineteenth-century ethologist George Romanes gave an account of two terns showing loyalty to a downed fellow tern wounded by a collector's shot. The rescuers held their comrade's wingtips in their beaks and flew him to a nearby island. A century later, three letters to the editor published in the August 2005 issue of the British magazine *BIRDS* provide comparable accounts:

> A dunnock hit a window and died instantly. Two others approached its body and dragged it a short distance to a shallow channel of water. One repeatedly lifted its neck in its bill as if trying to revive it. (Jim Hannan, by email)

> ... a blue tit hit my patio door and fell on the ground. Instantly three more arrived, calling hopping around the victim and nudging its head. ... [Then] one blue tit promptly flew down and butted the other's head: the stunned one righted itself but stayed motionless for some

> time. After more rapid head-butting, suddenly both flew away. (Mrs F Bailey, Kent)
>
> A blue tit hit a window and lay stunned underneath. Another flew down and crouched next to the injured bird; after about two minutes the new arrival suddenly violently attacked the injured tit, throwing it around the patio for about 30 seconds. Then the fit bird moved in close to the sufferer and propped it up with a wing until, quite suddenly, they both flew away. (Professor Alan Dyer, Lancashire)

Loving companionship often inspires relief or joy following an absence. If you've ever seen the unbridled outpourings of a dog reunited with a human companion you'll have seen a masterful blend of joy and supplication. American humorist and lecturer Josh Billings surely had it right when he quipped that dogs are the only things on Earth that love you more than you love yourself. When two infant chimps meet after a separation, they typically embrace, tickle and wrestle.

When Duane Callahan and Susan Marfield returned from a six month absence, the eight-year-old raven they'd raised flew immediately to Duane's shoulder. Merlin clung to Duane like a burr most of the day.

The intensity of parrot love

There are myriad anecdotes of expressions of love between different animal species. Not surprisingly, most published accounts of trans-species love involve humans. Animal love towards humans may be expressed in a variety of ways, including affection, gratitude, loyalty, jealousy and even sympathy, as when the chimpanzee named Lucy would run and hug and kiss her adoptive human parents Maurice or Jane Temerlin when either was sick and vomiting.

We may expect especially strong bonds in species with long lives and who mate for life. Parrots fit these criteria: they live for decades and usually stay with their mates 'til death. As popular companions, there are many accounts of their loving attachments with human caregivers. Captive parrots also bond with dogs, and will follow them around, even riding on their backs. Such is their loyalty to a chosen partner that parrots will typically reject a member of their own species once 'betrothed' to a human.

About five years after Joanna Burger adopted Tiko, the middle-aged parrot quite abruptly began courting her. When Burger accepted his courtship entreaties, she writes of being able to 'tell by his soft cooing that he was clearly in ecstasy.' That it took five years for Tiko's feelings towards Joanna to gel suggests this was no casual fling for him, but rather an emotional attachment built on years of developing trust and familiarity. That the object of Tiko's affection was not a member of his species suggests that sex provided little motive for his behavior.

Tiko expressed his affections for Joanna in various ways. He nipped her lips gently with his beak, as he would the beak of a female parrot. He also solicited preening from her, putting his head flat against the desk at which she was working, bill down, exposing his nape – as a gesture of complete vulnerability and trust.

Love can breed jealousy, and parrots are renowned for both. Tiko showed jealousy towards Burger on numerous occasions. One of the more dramatic periods was when Burger was nursing Hester, a hen, back to health after she was mauled by a dog. When Burger came home to find Hester had fashioned a nest using shredded newspaper in the sink of the bathroom where she was quarantined, she stood crooning to the hen. Witnessing this, Tiko careened onto Burger's shoulder and pecked her hair ferociously. It took 15 minutes of Joanna preening Tiko before he calmed down, and he kept Burger in his sight after that. Jealousy is hardly a pleasurable emotion, but its expression signifies a depth of emotional attachment akin to love.

Tiko spent long periods 'preening' Burger's hair and toes. He worked diligently and gently to remove a bandage covering a

Figure 8.2 Sammy, a salmon-crested cockatoo, leans in for a kiss with psychologist Lorin Lindner.

blister on her toe. When Burger was bed-ridden for weeks recovering from Lyme Disease, Tiko's ministrations reached new heights. He laid out every strand of Burger's hair in a fan-shape on the bed covers, moving Burger to write:

> As he cared for my body, I felt myself transported into a much more physically attentive kind of life than we're used to in this society.

Burger likened their relationship during this time to that of

> a mother who every night, in a ritual full of meaning and pleasure for both, brushes her daughter's hair.

In turn, Tiko was delighted when Burger proofread, because she then had one hand free to stroke him.

When Burger recovered from her illness to the point that she arose from bed and went into the kitchen to eat for the first time in weeks, Tiko seemed to celebrate. He swooped through the house, chattering, sliding down the banister, 'gleefully' throwing food on the floor, duetting with Joanna's husband Mike, and gobbling pinyon nuts, a favorite food that he'd forsaken during the illness.

The devotion of a parrot for his or her life companion is such that the bird will risk death in the defense of that mate. On the day that I finished reading Burger's book about her relationship with Tiko, the inside cover of the *Washington Post* reported that 'Bird,' a white-crested cockatoo, had helped produce evidence leading to the arrest of the two men who murdered his human companion, Kevin Butler. While Butler's assailants were stabbing the victim with knives, Bird landed on one of the killer's heads and pecked furiously. Bird was stabbed and killed with a fork, but blood from the man's scalp left at the scene provided the DNA that led to his arrest.

Conclusion

Because it belongs in the realm of emotions, love is difficult to describe in humans and even harder to interpret in an animal. It is not a pleasure that scientists are normally willing to entertain. And for that reason there is very little experimental data to shed light on this aspect of animal existence, and one is left having to speculate from anecdotes. Practically all animals must have sex to reproduce, and for many the choice of a mate is a serious business, especially when it's for the long haul. For these species, emotional feelings appear necessary for the sexual dough to rise.

Chapter 9

TRANSCENDENT PLEASURES

Humor, esthetics and beyond

Animals, like us, have rich and spacious interiors. They contain inner landscapes: desert places and lonely canyons, cliffs of madness and rivers of serene awareness that merge in tranquil seas.

Gary Kowalski, *The Souls of Animals*

The world is richly complex, and there are many ways not yet discussed from which animals might get a kick – drugs, esthetics, music, and humor, to name but a few. Most of us manage to steer clear of those illicit drugs such as cocaine, heroin, and amphetamines whose addictive potency can destroy lives, but will have had enough experience with alcohol or coffee to know the feeling, however temporary, of a drug-induced euphoria. We are all familiar with the joy of beauty, music, and humor. That we can relate to these experiences is significant because it is a grounding point from which we may appreciate other animals' experiences of similar stimuli.

Getting high

Workers at the Bronx Zoo recently devised a clever method for collecting the hair of wild cheetahs for DNA analysis. They fastened dog brushes to logs and put various scents on them to see what might attract the cheetahs to rub against them to leave hair on the brush. They tried a number of different scents, with varying success. Then they hit on Calvin Klein's *Obsession for Men*. The euphoric female cheetahs rubbed all over the *Obsession*-doused brush.

The domestic cat's response to the herb catnip is a more familiar variation on this theme. Animals crave it and will meow and climb to get at it. Under its spell, cats chase and paw at nonexistent mice and phantom butterflies. The plant contains nepetalactone, a chemical compound akin to a pheromone in the urine of sexually receptive cats. The solicitous rolling about of female cats under catnip's influence suggests that sexual arousal is a side effect. A cat who has discovered a catnip plant will return to it daily. Ditto cougars, lions, jaguars and leopards.

Cats on catnip are an excellent example of pleasure for its own sake. There is nothing to be gained from catnip with any survival value to a cat. The chemical basis of catnip's allure has evolutionary origins – i.e. procreation. Fun things often trigger a circuit evolved to do something else. There's rarely something pleasurable for us that doesn't have an evolutionary bedrock – music and art tickle our pattern-recognition software, and drugs trip the reward pathways honed to keep us mating and eating. Catnip is apparently simply fun for a cat – just as cannabis or tobacco are for us in tripping our reward switches.

Nature provides no examples of rhinos sniffing glue or lions elbowing up to the bar, yet there is plenty of evidence that many animals like getting stoned. Reindeer are partial to the hallucinogenic fly agaric mushroom. Jake, an enormous, placid bull, loved to sniff the fumes from the exhaust pipe of cattle farmer Rosamund Young's Land Rover. Elephants come running for the fermenting fruit of the marula tree, which they can detect from 10 kilometers. Under its alcoholic spell, they will throw the fruit at each other, and generally behave rowdily.

Among bird species known for gorging on fermenting fruits or berries, then showing the effects of their alcoholic stupor, are waxwings and robins in North America, and lorikeets and cockatoos in Australia. Cedar waxwings have a taste for fermenting rowan berries. Heaps of the birds have been seen dead beneath these bushes and post mortem examinations show they were drunk when they died and that they had acute alcoholic liver disease. Such behavior is not likely adaptive, just as alcoholism is not in humans. Merlin, the domesticated raven, sips beer offered him from an open can, and Duane Callahan describes him getting a little unsteady on his legs and wings on hot days, when he drinks a bit more than he should. Even insects are not immune to the occasional tipple. Bees and wasps are known to get 'drunk' on fermenting apples, which impairs their locomotion and their learning ability.

It is common for animals to return repeatedly to the source of a new-found plant intoxicant. Sometimes the consequences are disastrous, as when cattle develop a taste for fatal locoweed, or when bighorn sheep grind their teeth to useless nubs scraping a hallucinogenic lichen off ledge rock. Goats may deserve credit for the discovery of coffee: Abyssinian herders in the 10th century noticed their animals would become particularly frisky after nibbling the shrub's bright red berries. Pigeons spacing out on cannabis seeds may have led to this plant's use as a recreational drug by humans. Tukano Indians report jaguars – not ordinarily herbivorous – eating the bark of the yaje vine, then hallucinating.

Madagascarian lemurs and South American capuchins gently bite the bodies of large millipedes and rub them across their skin. The millipedes exude powerful defensive chemicals, including benzoquinone and hydrogen cyanide, and this fumigation may help ward off mosquitoes and deter external parasites. The primates also seem to enter a drugged state. They drool copiously and their eyes glaze over. Small groups may pass around a millipede like a marijuana joint, as they sink into a narcotic trance for 20 minutes or so before dropping the 'spent' invertebrate to the ground below, usually unhurt. There are obvious parallels here to recreational drug use in humans; describing it in his book *Weird Nature,* John Downer writes: 'the evident pleasure must act as some kind of reward.'

There are other variations on this theme. Many species of bird go in for a strange behavior called anting, or ant-bathing. A bird stands on an anthill, plucks the swarming ants in his beak and swipes them across his feathers. It looks much like preening, except the bird appears intoxicated, presumably by the potent formic acid exuded by the agitated ants. Some birds also smoke-bathe, sitting atop chimneys in apparent bliss, letting the smoke waft up through their spread feathers.

Intoxication is hazardous. Animals who get high on plants tend to be more accident prone, more vulnerable to predators, and less likely to attend to their offspring. In short, it can be maladaptive, which makes something of a mystery why animals are attracted to it in the first place. It is a fallacy that everything an individual animal does confers evolutionary benefit. That animals abuse drugs in nature is evidence that they, like us, will sometimes behave badly simply for kicks.

This is not to deny advantages to certain psychoactive compounds. They may relieve pain, enhance concentration, extend endurance, sharpen eyesight, increase strength, unleash inhibition, relieve stress, induce or repel sleep, rouse or quell aggression, or stir the sex drive. The use of medicinal plants by animals is so widespread that there is even a term for this field of study: zoopharmacognosy. Usually animals partake of plant compounds not so much to experience pleasure as to relieve aches and pains, and there are many examples of animals consuming bitter-tasting leaves, seeds and flowers to quell painful ailments and discomforts, including digestive parasites and menstrual pains.

Mad with joy

The *Oxford Pocket American Dictionary* defines joy as 'a vivid emotion of pleasure; extreme gladness.' To most of us it is usually a spontaneous burst of delight and good feeling, such as that following a surprise or a long anticipated event, such as a reunion with long-parted friends or family.

Without a fitting context, it might be difficult to identify many instances of animal joy. But when a situation befits the response, animal joy is unmistakable. When two female chimpanzees were unexpectedly confronted with an enormous pile of bananas, 'they flung their arms around each other's neck and pressed their open mouths to each other's shoulders while uttering excited food calls before they took a single fruit.' Similar behavior has been spotted in rhinos, elephants, cats, dogs, gorillas, bears, horses, orangutans, spinner dolphins, beavers, mountain goats, chamois, chimps and bottlenosed dolphins. Liberation from dreary confinement produces expressions of joy in various animal species. When the chimpanzees leave their indoor winter quarters at the captive colony at Arnhem Zoo, The Netherlands, chimpanzee expert Frans de Waal describes it as 'the most festive day of the year':

> ... All over the enclosure apes can be seen embracing and kissing each other. Sometimes they stand in groups of

> three or more jumping and thumping each other excitedly on the back. The apes' delight in regaining their freedom is obvious. Their black coats, which have grown thin during winter, will become thick and shiny again within a few months. Pale faces will regain their color in the sun. And, most important of all, the tension, which has been bottled up all winter, will dissolve again in the open air.

Like us, chimps are not particularly fond of being out in the rain. When primatologist Wolfgang Köhler – studying chimps at a field station in north Africa around 1915 – went out in the rain to let in two chimps locked out of their artificial den, he recalls that, before entering, the grateful chimps 'turned to me and put their arms around me, one round my body, the other round my knees, in a frenzy of joy.'

Charles Darwin wrote of his good fortune of happening to visit the Zoological Society of London on the day that the rhino was let outside of his indoor enclosure for the first time of the year:

> Such a sight has seldom been seen, as to behold the rhinoceros kicking and rearing (though neither end reached any great height) out of joy.

A few years after Darwin visited the London zoo, the American novelist Stephen Crane was touring a coal mine near Scranton, Pennsylvania. Though his main intent was to describe the conditions endured by human miners, he was moved also to note the behavior of the mules used to pull the coal and slurry. For convenience, the animals were kept in the mine for years at a time. Crane describes a rare return to the surface for the animals:

> the mules tremble at the earth radiant in the sunshine. Later, they almost go mad with fantastic joy. The full splendor of the heavens, the grass, the trees, the breezes,

> breaks upon them suddenly. They caper and career with extravagant mulish glee.

Spotted dolphins escaping from the purse seine nets lain out by fishing vessels to catch tuna (which commonly school beneath the mammals) leap repeatedly into the air, as though in joyful celebration. Cattle, when first let out into the fields following a long winter confinement, tear about the field, kicking their legs into the air. They seem literally to be full of the joy of spring, and look for all the world like excited toddlers released into the playground after hours sat behind their desks.

Ethologist Joyce Poole, who directs the Amboseli Elephant Research Project in Kenya, describes elephant reunions – following hours, days or weeks apart – as nothing short of pandemonium: members rush together, ears flapping and heads high. They urinate and defecate while they spin around, and secretions flow profusely from their temporal glands. Low-frequency rumbles, screams, roars and trumpets cleave the air. Mired in a rut of adaptive significance, we may look only for survival advantages in such outpourings, for example that they send a message to neighboring elephant groups that they are a force to be reckoned with. Poole believes that the ruckus is best interpreted as 'Wow! It's simply fantastic to be with you again!'

Joy can certainly be communicated across the species barrier, as practically anyone knows who takes their dog for walks. When I was young we learned to stop mentioning the word 'walk' when it was time to take our little mongrel dog, Begs, outside. At first we took up spelling it, but Begs soon caught on to the sound of 'double-you ay el kay' and we just had to put up with the chaos produced by a small furred projectile shooting through doorways and ricocheting off shins and furniture.

Rats, especially young ones, chirp with apparent joy when playing and in anticipation of play. These chirps are in the 50 kHz pitch range (well above our upper hearing limit). Rats make the same 50 kHz chirp when anticipating various treats, including narcotic drugs. These chirps are quite distinct from the sounds they make when negative events occur – a long 'complaint' call at 22 kHz. Rat chirping may even be a primitive kind of laughter. When tickled on the nape, where play is usually initiated, they utter many chirps. In a controlled experiment, rats unambiguously chose to spend time with rats who chirp (50 kHz) a lot rather than with those who don't.

The joy of flight

Flight appears to be a source of joy for many birds. There are innumerable sightings of birds behaving as though flight is so much more than a means of transit, showing off, defending or foraging, as we saw in the chapter on play. One clue to recognizing joyous or playful bird flight is how it contrasts with serious, business-like flight. In the colder months when one can watch scores of crows making their daily aerial commutes between Washington, DC, and the Maryland suburbs (where I make my permanent home), they fly unwaveringly straight ahead and at a constant speed. Their flight pattern contrasts with the relatively erratic, rolling flight one often sees in crows. If crows are capable of flying straight and true, getting themselves to some intended destination in the most efficient way they can, then their 'flights of fancy' might sometimes reveal their lighter side, and show that these birds are not all business.

Large, intelligent and easily watched, crows and their ilk (ravens, jackdaws, magpies, and jays) offer many opportunities to witness flight for flight's sake. On a sunny late afternoon in northern England in September 2003 I watched a small flock of carrion crows foraging on a grassy slope. It was windy, and at intervals, one to three birds would take wing, bounce, slide or swoop into the air currents, then land again. There was no apparent purpose to these little flights except to experience the joy of it. I have also watched crows flying around the topmost tower of England's famed York Minster cathedral, repeatedly diving and circling in swoops around its girth. The character of their flight was frolicsome, with a distinct air (if I may) of joy and exhilaration. Alas, rare is the human down below who stops to notice, or to wonder if birds are having fun.

On a visit to South Stack on the coast of Wales in 2004, I watched a lesser black-backed gull who dipped and swerved into the updraft at the crest of a cliff-face for 10 minutes. At times, she or he would drift out into the ferocious headwind, then double back to resume the game. The bird's behavior was purposeless (unless you take the reasonable position that fun is an end in itself) and her movement buoyant, like a ping-pong ball kept aloft on the column of air blown through a straw.

Gulls and crows are not the only sorts of birds seen to ride coastal winds. In his 1952 book *The Personality of Animals*, H. M. Fox describes

> ... hundreds of razorbills and puffins together, flying over the sea just beyond cliffs, in a gale of wind. They flew round and round by the hour, in a great ellipse half a mile across – a splendid sight. What was the purpose of this flight? One could guess at none.

On a crisp sunny February afternoon near York, in northern England, I watched a flock of 50 jackdaws flying about 70 m overhead. They circled in clusters, giving out their distinctive upslurred call. As the flock drifted downwind, it began to disintegrate as individuals dived earthwards. Many swooped as they approached the ground at high speed. The spectacle had an air of *joie de vivre* about it. Mark Brazil, who writes a regular natural history column for the *Japan Times*, reported a group of 32 ravens engaged in aerial pursuits, paired flights, swoops, stalls, and rolling displays. Among them, two individuals flew towards each other, grasped each other's beak and descended slowly for several seconds, their wings and tail feathers extended like a black parachute.

Is it evolutionarily advantageous for birds to enjoy flight? Perhaps, but that is not to suggest that pleasure drove the evolution of flight. The evolution of flight reaped more salient benefits: efficient foraging, escaping predators, and the exploitation of profitable niches, such as aerial insect-eating. Thus, while adaptive living and pleasure are synergistic, it would be erroneous and teleological to think that flight evolved in birds because it was fun. Pleasure arises secondarily, from the reward systems that natural selection encourages. Because flight is also costly in terms of energy required to develop wings (and to use them), birds evolving on islands where there is plenty of easily available food and no predators tend to be flightless. In the absence of selection pressures favoring flight, birds who continued to fly because it was fun might be at a disadvantage compared to birds who stayed on the ground and conserved energy.

The preceding discussion is heavy on anecdote and light on rigor. This need not undermine the notion that birds may derive joyous pleasure from flying. That there is so little science to bear on the subject is partly because it is so difficult to study in any sort of systematic way. If birds do enjoy flight, its source of pleasure likely derives from its freedom and spontaneity. Flight joy probably requires open spaces, and the sort of well-being – dare I say, happiness – that would be hard to cultivate in a controlled, captive environment. Nor should we assume that if one species of bird derives pleasure from flying, then all birds do. Excluding corvids (crows and their allies) and larids (gulls) from my own observations, there's very little left. What about all those little birds – the finches, thrushes, wrens etc. – who seem always to be so serious and businesslike in their flight activities? Can a male skylark revel in his incredible courtship flight, which ascends to hundreds of feet and is accompanied by an outpouring of song that may last for 15 minutes, before he parachutes, then nose-dives back to his pastoral patch? Or are we, his land-bound observers, the only ones feeling any exultation? I often glimpsed high-speed aerial chases among the swifts who roost in a large 19th-century church across the road from my home in York. Swifts commonly chase one another, and the behavior has been described in at least one scientific paper, but it isn't known whether they involve positive (e.g. playful) or negative (antagonistic) feelings in the participants.

Practically all of my own observations, described above, have occurred since I conceived the idea for this book. I'd been watching birds for 30 years before that, but I never gave more than a moment's thought to how they might experience it, beyond my own envy for them. That I've since noticed so much more in the behavior of flying birds that suggests some pleasure in it testifies to the importance of an observer's attitude and expectation. Obviously, we should be wary of reading too much into the apparent emotive character of creatures which cannot tell us directly how they feel. But we may stand to gain insights into the more elusive aspects of animal existence when we approach then with a genuine, if guarded, degree of empathy.

Esthetics

Do animals have a sense of beauty? Do they appreciate esthetics? A cursory glance at nature leaves scarce doubt that they do.

The plumage of birds suggests considerable overlap between what humans find visually pleasing and what birds do. Males of many bird species sport bright colors during their breeding season, advertising their beauty and fitness as mating partners. Female choice has played a major role in the evolution of bird mating systems, and the displays of peacocks, pheasants and birds of paradise testify to the females' appreciation of esthetic beauty. Female blue tits, for example, prefer males with crests that look brightest in the ultraviolet spectrum. Among vertebrates, birds and fishes are the best endowed for seeing colors; their eyes contain four types of photopigment, whereas those of humans and other primates contain three. As evolutionary ethologist Rick Prum puts it: 'Birds look great to us, but they look a lot better to themselves.'

Some of the pheasants have, quite literally, taken these esthetic displays to great lengths. Male Temminck's tragopans can

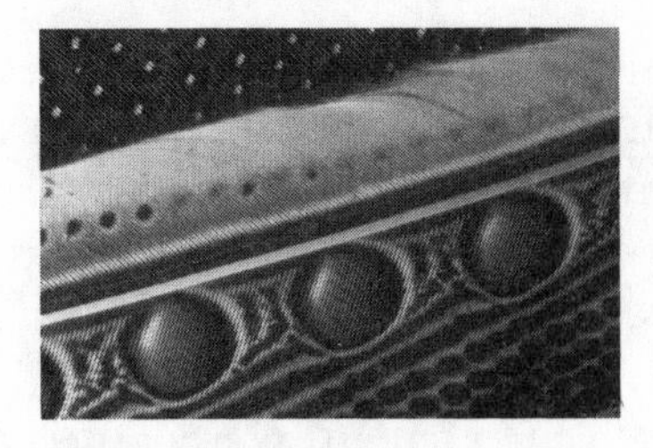

Figure 9.1 A beautiful argus pheasant feather.

pump blood into a large throat wattle, expanding it into a spectacular blue and crimson bib. The male monal pheasant displays a huge expanse of unbroken iridescence on his neck, wings and tail. Argus pheasants, like their better known peacock cousins, erect a gorgeous fan of long tail feathers, each with simulated 3D spots. These ornaments come at a price. They are cumbersome and energetically expensive to produce, maintain and carry around. They also make the wearer more visible to predators, and less able to escape. That they evolved in such splendor illustrates the power of esthetics in nature. (Another price has been our own appreciation for their beauty, leading to their near extinction. Egrets and spoonbills, for instance, practically disappeared from the Florida everglades before a law was enacted in the 20th century protecting them from hunters supplying a booming millinery market in Europe.)

Some birds' visual tastes are expressed by more than the costumes nature has bequeathed them. Courting male and female cedar waxwings may pass flower petals back and forth as if exchanging gifts. The selective use of petals suggests an esthetic element, otherwise why not use leaves, which are more readily obtained but drabber? Bowerbirds are renowned for the elaborate nest-like structures they build. These bowers serve solely to attract a female, for she builds a separate nest to lay eggs and raise her chicks. Vogelkopf bowerbirds produce especially elegant galleries to attract females. Among the *objets d'art* they accumulate for their showrooms are flower petals, fruit, beetle elytra (wing coverings) and stones, each carefully arranged into neat piles or arrays. Studies show that male attractiveness to females hinges on several factors, including bower decorations and bower quality as well as the intensity and quality of a male's courtship.

In his studies of bowerbirds, German biologist and film-maker Heinz Sielmann described how they inspect and refine their bowers, likening it to visual artists who assess the progress of their work. On adding a decoration, the bird moves back to a proper distance for surveying it, then may make adjustments if it doesn't quite have the desired effect. 'Painting with flowers is the only way I can put it,' said Sielmann. Some bowerbirds even paint parts of their bowers with charcoal and crushed berries, using a bark 'brush.'

Studies have shown that pigeons can differentiate the paintings of Monet from those of Picasso. Again, this does not necessarily indicate an esthetic appreciation for the artwork, but it does indicate an acute sense of observation, not just for color and detail, but for nuance and gestalt. Perhaps the birds are cueing in on elements of a painting – such as the subtle differences in brush-stroke character – to which only the most trained art critics are attuned.

Plants display almost inexhaustible variations on beauty. A stroll through a botanical garden is an excellent way to appreciate this. I visited Copenhagen's Botanical Gardens, a green space whose stately charm befits this venerable European city. Small birds flitted about beneath the protective cover of exotic cedars and pines, ducks and moorhens paddled quietly in the shallows of a central pond strewn with lilies, and a magnificent chain of circular greenhouses rose above the morning mists.

With each new flower, I encountered a variant on the theme of sensory attraction. This was an art gallery with nature as artist. Given hundreds of millions of years to work on her canvas, she has summoned a stunning array of enticements. Colors ranged from subtle to intense, some monochromatic, some with one hue blending gradually into another, others forming contrasting patterns. Shapes were as diverse as pigments: simple arrays of flat petals, tubes with fuzzy platforms, wrinkles, whorls, and tall spikes encircled by a single ring of blooms part-way up were among a seemingly endless variety. Scents spanned the undetectable to the pungent to the intoxicating. Even the scientific names were beautiful – *Dioscorea bulbifera, Hedyscepe canterburyana, Elingamita johnsonii, Andredera cordifolia, Stromanthe sanguinea, Dioon edule* – and I could imagine a Victorian botanist in a cluttered, oak-paneled room leafing through a musty Latin dictionary to find a term that captures the character of a newly discovered plant.

Drawn into such a multi-sensory feast, it is easy to forget that all this magnetism is not directed at us. Most flowering plants graced the earth long before hominids descended from the trees. Few if any flowers have evolved solely to be pollinated by humans. No, as poet and critic Frederick Turner muses in his book *Beauty*, these jewels germinated for other beasts:

> The colors and shapes of the flowers are a precise record of what bees find attractive.... It would be a paradoxically anthropocentric mistake to assume that, because bees are more primitive organisms ... there is nothing in common between our pleasure in flowers and theirs.

To bees, we can add wasps, beetles, flies, butterflies, moths, fruit bats, and many birds, including most notably the hummingbirds and sunbirds.

Symmetry in plants is something of an extravagance. While the need to move in a straight line has favored the evolution of symmetry in animals, there is no such evolutionary pressure on plants. Symmetry must confer benefit. Presumably, it has evolved in flowers because it appeals to pollinators. It makes different flower species more distinctive and recognizable than if they were more randomly shaped. Clover pollen is of no use to a

Figure 9.2 Exquisite flowers evolved long before humans were around to appreciate them.

daisy, so helping insects recognize and target a specific flower species benefits reproduction in that species.

Our esthetic appreciation of pattern, symmetry, and color – widely evident in architecture, fashion, and art – is not some arbitrary artifact of human culture. Its roots run deep in the complex interdependencies of the living world.

Music

Music forms an enormous part of human culture. It is a multi-billion dollar per year industry based on sensory pleasure. The sounds we appreciate vary hugely over time and place; the same goes for other animals, from the melodic songs of birds and the percussive clicks of fishes to the buzzes and trills of insects and the resonating anthems of whales.

We would be naïve to think that our music sprang spontaneously from our cultures, independent of the sound that already filled the skies (and seas). So too would it be narrow thinking to assume that only we derive pleasure from sound. Sound is an important medium of communication in the animal kingdom. If you are outdoors or near an open window as you are reading this, chances are you may be within earshot of some form of animal sound. As I write in an English loft, I can hear chirps and chatters of house sparrows, a barking dog, the cooing of a rock dove from a nearby rooftop, and occasional snatches of blackbird song.

Birds' songs perhaps come closest to the tuneful forms of music to which we gravitate. Ethologist Colin Beer notes how harmonic content, melodic themes and rhythms of formal musical organization derive from an in-born esthetic sense. Alexander Skutch, in his book *The Minds of Birds*, remarks on how the tonal qualities of bird calls tend to resemble those we use in similar situations: lullabies on the nest, harsh or grating sounds when warding off predators, mournful cries when watching an intruder at the nest. Each suggests that birds' emotions share origins with our own.

Len Howard, an English musicologist, devoted more than a decade of her life to the study of the wild birds that lived in her house during the 1940s. She refitted the interior to provide perching and nesting places, then left the windows open during the spring and summer months. Many birds took up residence, built nests and reared young. Among the conclusions she made from her intimate study of birdsong was that in addition to singing with definite goals in mind (territory, mate attraction), birds also sing simply because they are happy.

Noted philosopher and ornithologist Charles Hartshorne felt much the same way. He recorded and analyzed the songs of hundreds of bird species over fifty years, and concluded in his 1973 book *Born to Sing* that birds sometimes sing because they are happy. Hartshorne was not afraid to celebrate birds' experience of song, and he recognized the compatibility of pleasure and survival:

> There is no conflict between 'birds sing for pleasure' and 'they sing to maintain territory or attract mates.' The more essential an activity in the whole life of the bird, the greater the proportion of the bird's pleasure which is realized in that activity.

A lifetime of bird observation convinced Hartshorne that birds sing for social and psychological reasons. He wrote:

> the appeal of music is, first of all, sensory and emotional, not intellectual. Far more than we do, birds live in a world of sensation, feeling, and impulse.

The feelings we may have in response to the sounds of other animals are indicative of our biological connection with them. Could it be that the wide appeal of birdsong has a biological basis? An emerging field, called biomusicology, explores the idea of a 'universal music,' an intuitive musical concept shared by many animals, including humans. The idea that other species may share an innate biological musicality with us is appealing, if a little hard to assess. Even one's own response to a particular bird's song is personal, and not easily described. Many, though, have been inspired to try. American musicologist and humorist P. D. Q. Bach was so entranced by the tunes of a wood thrush emanating from the woods near his home that he composed a piece of music to it. Mozart rewrote a passage from the last movement of his Piano Concerto in G Major to match the song of his beloved pet starling. The solo piano piece 'A Hermit Thrush at Eve,' by the American composer Amy Beach, was the result of an extensive musical dialog between the thrush, who sang his 'lonely and appealing' songs over and over, and the composer, who wrote them down, revising with each repetition.

Whale expert Roger Payne described humpback whale song as a 'vast and joyous chorus of sounds ... sounds that boomed, echoed, swelled, and vanished as they wove together like strands in some entangled web of glorious sound ... sheer ebullience ... lovely, dancing, yodeling cries.' The feelings expressed in this poetic passage are those of the listener and not (necessarily) the performer. Nonetheless, we should not dismiss the biological continuity between whale and human.

Research by Payne and his wife led to the discovery that humpback whales' songs may be more than an hour long, and in a given year, all individuals in the Atlantic Ocean sing a similar song, as do those in the Pacific. These songs string together subphrases into phrases, phrases into themes, and themes into songs; a session of whale songs can go for 10 hours.

While there's no question that humans appreciate the beauty of animal music, there is some evidence that the reverse is also true. Joanna Burger's companion parrot Tiko enjoys music. He has a particular liking for Baroque harpsichord, and Scarlatti is a favorite. Joanna's ethologist husband Mike whistles duets with Tiko, who seamlessly repeats Mike's phrases such that Joanna can hardly distinguish them.

Several studies have shown that birds recognize and show preferences for human music. Java finches can tell apart Bach and Schönberg played on the same instrument by the same performer. Trained birds could correctly identify pieces by these composers which they had not previously heard. They could also generalize to different composers, linking Bach to his contemporary Vivaldi, and Schönberg to another modern composer, Eliot Carter. Some of these birds preferred the early music, some did not. Pigeons have also been found to generalize from Bach to Buxtehude and Scarlatti, all Baroque composers, and among the modern composers Stravinski, Carter and Walter Piston. European starlings can recognize different chord types and rhythms.

Mammals also show evidence of musical sensitivity. Rats performed better in a maze after being exposed to Mozart compared with Philip Glass, and studies find that cows' milk production is higher with slow than with fast music.

These results indicate that birds and mammals have individual preferences for certain types of music, and suggest that they might perceive musical stimuli much as we do. It is reasonable to hazard, therefore, that sound may have esthetic value for them, too.

Laughter and humor

For a long time laughter has been included on a dwindling list of things uniquely human. It now appears some animals are having the last laugh on this one. Or perhaps more accurately, the first laugh. Laughter has instinctive elements. It arises from deep in our animal nature, and involves the amygdala and hippocampus, two structures in the ancient limbic system of the brain. Research suggests that the capacity for human laughter preceded that of speech during brain evolution.

Scientists have known for a century that the great apes display something similar to human laughter. Darwin noted:

> We may confidently believe that laughter, as a sign of pleasure or enjoyment, was practised by our progenitors long before they deserved to be called human; for very many kinds of monkeys, when pleased, utter a reiterated sound, clearly analogous to our laughter, often accompanied by vibratory movements of their jaws or lips, with the corners of the mouth drawn backwards and upwards, by the wrinkling of the cheeks, and even by the brightening of the eyes.

Darwin also noted that young chimps on being tickled, especially under the armpits, behave much like humans do when we chuckle or laugh, though it is sometimes noiseless. They open their mouths wide, expose their teeth, retract the corners of their lips, and utter loud, repetitive cries much like ha!ha!ha! Frans de Waal notes that a panting sound accompanying chimp wrestling and tickling games is a lot like 'suppressed laughter.'

Chimps laugh in many of the sorts of situations where we would laugh: during tickling, rough-and-tumble play, and chasing games (the one being chased typically laughs most). Their vocal output is about twice the speed of ours because their sonic bursts are produced during both inhalations and exhalations. So, unlike us, they don't lose their breath during their 'panting laughter.' Chimps will laugh alone, such as when playing with objects, tickling their own feet, rolling on the ground, and swinging. And like humans, chimps laugh more when they've quaffed alcohol.

All of the great apes – humans, chimps, bonobos, gorillas, and orangutans – share the tickle spot under the arms. In her early studies of the chimps of Gombe, Jane Goodall described playful interactions between the adult males Goliath and David Graybeard, which began with hand tickling, then body tickling, tumbling and 'laughing,' then chasing, followed by a 21-minute grooming session.

Michael, one of the gorillas at the Gorilla Foundation, plays a tugging and chewing game with his caregiver through the bars of his enclosure. Michael occasionally gets what we might refer to as 'the giggles,' and may laugh continuously for up to 10 minutes during this game. Laughter's contagious quality extends to gorillas; once he and DeeAnn his human teacher get laughing they 'feed off each other's joyous mood for quite some time.'

Jaak Panksepp has been studying what he calls laughter in tamed rats during rough-and-tumble play. Rats chirp most robustly during play and tickling; like us, they have ticklish areas, particularly around the nape of the neck. The chirps are not an artifact of thoracic compression during rough-and-tumble play, as some skeptics have suggested. Response to tickling also declines with age, which Panksepp sees as consonant with the observation that human children are more ticklish than adults.

Patricia Simonet of Sierra Nevada College in Lake Tahoe has been studying a sound that dogs make while playing that spurs others to play when they hear it. Though the sound is very different from human laughter, the researchers believe it is the canine equivalent. It comprises a sharp burst of sound containing many more frequencies than a dog's plain pant. Puppies as young as eight weeks old become playful when they hear a recording of this dog 'laugh,' something they don't do when presented with other dog sounds.

Panksepp believes that rats and primates, especially youngsters, might use laughter to signal playful as opposed to threatening physical interactions. Play-fighting puppies tumble over each other, snarl, and pretend to bite or give chase. But even we can tell, by aspects of their behavior, including their vocalizations and the intensity of their movements, that the fight is fun and not serious. It is not uncommon for the intensity of play to escalate to the point that the line between 'fun' and 'serious' becomes blurred; I have seen this many times when dogs play boisterously. A puppy's inability to signal (because of a damaged brain or having been raised in isolation) can lead to serious fighting.

If laughter has been around for long before humans evolved, then we may expect that it has some survival benefit. It regularly does in humans. Research is showing that laughter decreases stress and anxiety, reinforces immunity, relaxes muscle tension, and decreases blood pressure and pain. The 1998 film *Patch Adams*, starring Robin Williams, chronicled the true-life work of a young doctor who uses humor to lift the spirits of seriously ill patients. Adams was scorned and reprimanded for his unorthodox ways, but today the Gesundheit Institute he founded in West Virginia applies these novel healing principles. Hospitalized children who see clown shows have shorter hospital stays than those who do not. Recent studies at the University of California at Irvine's College of Medicine showed that just looking forward to a good laugh is therapeutic. Merely anticipating seeing a funny movie or comedy act was found to lower cortisol, a stress hormone, and raise levels of pleasure-inducing endorphins.

That some animals laugh need not mean that they have a sense of humor. As science writer Eugene Linden correctly points out, 'the issue of humor, like the issue of lying, remains beyond the capabilities of rigorous testing because it is next to impossible to establish intent.' But by watching the behavior of other animals, we can at least gain insight into their potential for a sense of humor.

Koko the gorilla has been the source of many anecdotes suggestive of a sense of humor. When asked to identify the color of a white towel held up by a teacher, Koko signed 'red.' When asked again, she repeated this three times. Then, grinning, she plucked off a bit of red lint clinging to the towel, held it up to the trainer's face and signed 'red' again. She repeated this trick later with her primary teacher, Francine Patterson. (Interestingly Koko appears not to have repeated the trick with the same person, suggesting that she knew it would only 'work' once before the trick was discovered.)

Koko also makes plays on words. She has signed 'Koko + nut = coconut,' and put a straw to her nose, calling herself a 'thirsty elephant.' When Koko was a youngster and she was asked by her teachers to do something funny, she responded by feeding an M&M chocolate to a bird puppet in its eye rather than its mouth, and she put a toy key on her head and called it a 'hat.' These are instances of 'incongruity humor,' a form known from child development studies. Bongo, a gorilla at Columbus Zoo, 'would run along with keeper running outside, then stop suddenly and then laugh as the keeper whizzed past outside the bars.'

Anecdotes like these are not enough to conclude that gorillas have a sense of humor. But as they accumulate, it will become increasingly difficult for a nay-sayer to deny it without making others laugh.

We have all witnessed that brand of derisive humor unleashed at the expense of another. At Arnhem Zoo in the Netherlands, a disabled adult chimp named Krom was teased for several days by young chimps, who walked behind her mimicking her distorted, hunched-up gait. Dogs often wait until we are close before picking up a stick or ball and running a short distance away. Charles Darwin believed this teasing behavior was evidence of a dog's sense of humor. In *My Fine Feathered Friend*, the New York Times restaurant critic William Grimes describes how a chicken companion who would delicately trot up behind a particularly nervous cat named Yowzer when his back was turned, utter a loud squawk, and then cackle and run off jauntily as the cat shot two feet into the air.

Dolphin researcher Bernd Wursig has often seen dusky dolphins sneak carefully up on kelp gulls resting on the water surface, gently but firmly grab one or both of the bird's legs, then dunk briefly before letting go; the bird flutters, kicks, preens frenziedly, then flies off. Young captive orcas will loiter around the edge of a tank, which tends to lure people closer. They then drench them with a flick of their tail flukes. Cetaceans may not express laughter in ways that we would recognize, but this sort of mischievous play is consistent with a sense of humor. That it doesn't normally cause any serious harm to the butt of the joke also suggests a level of moral development. Parrots, both in the wild and in captivity, are also known to play games with others of their own kind that seem to serve only as a means of getting a reaction.

Conclusion

We are not crows, dolphins or chickens. We cannot experience a crow's feelings when he hurls himself off the high tower of a cathedral, an elephant's when she slurps up another round of fermenting marula juice, or a monal Pheasant's reaction to a male's beautiful display of plumage. But we can relate them to our own feelings in similar situations. And with careful, critical study, we may begin to draw some enticing conclusions.

Chapter 10

FROM FLIES TO FISH

At the margins of pleasure

> *A Chinese philosopher once had a dream that he was a butterfly. From that day on, he was never quite certain that he was not a butterfly, dreaming that he was a man.*
>
> Anonymous

Amphioxus is a small, unassuming fish-like creature. It would be relegated only to obscure ichthyology volumes were it not an important evolutionary link between those animals with and without a backbone. As such, most biology students at least learn its name.

Residing as it does at the evolutionary cusp between invertebrate and vertebrate evolution, *Amphioxus* is also an important symbol for the evolution of pleasure. For if there is any dividing line along which science might favor separating feeling from non-feeling animals, it is that which separates animals with and without backbones. Those creatures which reside in taxonomic proximity to *Amphioxus* might be regarded as the marginal cases. These include the fish, and all those animals lying on the other side of the great vertebrate divide – those without backbones.

It should be said of 'fish' that they are an enormously diverse group of highly evolved animals. They represent more species – 27,000 extant – than all of the other vertebrate classes (mammals, birds, reptiles and amphibians) combined. Fish have been evolving on Earth for about 400 million years, and for every type of fish that swims today, perhaps 10,000 species came and went before, based on estimates from fossil records and extinction rates. In any event, 'fish' is a collective term of convenience; what applies to one species may not for another.

On the other side of Backbone Avenue we have the invertebrates. This is a vastly more diverse collection of creatures. All the world's vertebrates are a village to the metropolis that is the invertebrates. Depending on which expert you consult, they comprise between 30 and 36 different phyla – each of which is a level of animal classification on par with the chordates (which contain all of the vertebrates plus a few anomalies like *Amphioxus*). Among the ranks of the spineless we find the hugely successful phylum arthropoda, which according to Ernst Mayr embraces some 4,650,000 described species (and counting), of which about four million are insects. Another large phylum, the nematodes (roundworms), numbers 500,000 species or so, of which only 20,000 have been described so far. Thus, for every vertebrate species alive today, there are about ten species of roundworm, and for every roundworm, about ten arthropods. Add to these big gatherings the mollusks (e.g. clams, octopuses, squids, snails, slugs – about 120,000 described species) and such lesser known phyla as the platyhelminthes (flatworms – 19,000 species), annelids (segmented and marine worms, and leeches – 15,000 species), and echinoderms (sea stars and sea urchins – 7,000 species), and you can see that we're dealing with a vast pageant of beasts. For reasons of space, and limited knowledge, I will be sampling from only a tiny subset here, primarily within the arthropods and mollusks.

So, preconceived notions aside, let's take a brief look at some of the evidence concerning whether or not fishes, and certain members of the invertebrate horde, might be able to feel good.

Getting a feel for fish

On the strength of their diversity, fishes have been called 'the most successful vertebrates.' Fishes are still commonly believed to lack feeling, to constitute unthinking creatures with no awareness, whose reactions to stimuli are merely reflexive. These dubious assumptions lean heavily on prejudice and convenience. Because fish have been in existence much longer than have so-called 'higher' animals such as monkeys and great apes, they are usually viewed as 'primitive' and less highly evolved. 'Primitive' is a value-laden and deceptive term; many fish retain ancient features because they work well, but it is not as if they stopped evolving. One could as soon claim that fish are more highly evolved, for they have been doing so longer. As the most ancient of the major vertebrate groups, fish have had ample time to develop complex, adaptable and diverse behavior patterns that rival those of other vertebrates.

Another reason why fish are commonly denied feeling is, I suspect, their relative lack of facial expression. When they are impaled on a hook, fish don't scream or grimace, though their gaping mouths may evoke a look of shock or horror to the empathetic witness. Using facial expression as a guide for sentience is hardly valid when one considers that some of the most intelligent and highly sentient marine vertebrates – namely the dolphins and whales – also lack facial expression, at least any that most of us can readily detect.

However, animals have many other ways of visually signaling their feelings. Crests, dewlaps, mouth-gapes, pupil dilation and contraction, color changes, and body postures and movements are among the many visual ways fish and other animals convey emotions. Water is also a potent medium for communicating via chemicals and sounds.

It has been hypothesized that fish and amphibians lack emotional and conscious experience based on their failure to show an emotional fever response. This is a measurable rise in an animal's core body temperature attributable to a psychological rather than physical cause. In humans, for example, emotional fever has been documented in students in the hours before their examinations, and in spectators at sporting events (especially if they themselves have competed at the sport). In rats, mere handling by an unfamiliar person elicits a rise in body temperature of 1° C or more. The elevation is considered emotional (and not, for instance, due to increased activity or transfer of heat from the human handler) because the response declines and disappears on repeated days as the handler becomes familiar to the rat, but returns if a new handler is introduced. Emotional fever has also been documented in birds, including chickens, and reptiles such as turtles and lizards.

The presence or absence of emotional fever (with a paucity of data to boot) is a narrow and tenuous basis for denying emotional or conscious experience to such vast groups of animals. In fishes at least, there are many alternative perspectives that support consciousness and emotion. And empirical studies are beginning to offer support. For instance, fish take 48 hours to return to normal hormonal levels following rough handling (e.g. being put into small buckets after fishing).

A recent study found that when noxious substances were applied to the lips of trout, the fishes' heart rates increased, and they took longer to resume feeding. These fish also exhibited unusual behaviors after being harmed, including rocking from side to side while balanced on their pectoral fins, and rubbing their lips into the gravel and against the tank walls. Treatment with a pain suppressant significantly lowered these effects. Other experiments have found that fish learn to avoid unpleasant stimuli such as electric shocks, and piercing of their lips by sharp hooks.

This behavioral evidence of pain perception should come as no surprise. In vertebrates, free nerve-endings register pain, and fish have these nerve-endings in abundance. Fish also produce enkephalins and endorphins, opiate-like substances known to counter pain in humans. When an animal responds to an unpleasant stimulus by directing attention to the injured part of its body, it suggests response to pain, not nervous reflex. The gains of pain perception, like avoiding danger and facilitating healing, apply as well to fish as to any other vertebrate animal.

As new evidence mounts, it is becoming increasingly difficult to support traditional views of fishes as dumb and numb. Writing in the journal *Fish and Fisheries*, biologists Calum Brown, Kevin Laland and Jens Krause make the following eye-opening statements:

> Gone (or at least obsolete) is the image of fish as drudging and dim-witted pea brains, driven largely by 'instinct,' with what little behavioral flexibility they possess being severely hampered by an infamous 'three-second memory.' ... Now, fish are regarded as steeped in social intelligence, pursuing Machiavellian strategies of manipulation, punishment and reconciliation, exhibiting stable cultural traditions, and co-operating to inspect predators and catch food.

Not so long ago, such claims would have been considered heretical when applied to monkeys, let alone fishes. Yet modern fish representatives recognize individual 'shoal mates,' acknowledge social prestige, track relationships, eavesdrop on others, use tools, build complex nests, and exhibit long-term memories.

Recent studies by Theresa Burt de Perera at Oxford University have shown that Australian crimson spotted rainbowfish, which had learned to escape from a net in their tank, remembered their technique 11 months later. In the same lab, blind Mexican cave fish – which rely on subtle changes in pressure to detect objects near them – took just a few hours to build a detailed mental map of obstacles that Perera placed in their tank, and quickly adjusted when she swapped obstacles around.

An amazing spatial memory ability has been discovered in the frillfin goby, which inhabits rocky tide-pools during low tide. If a pool begins to dry up, these fish leap to an adjacent pool. Obviously, a missed leap might be fatal, and the accuracy must be great in terms of both distance and direction. How do frillfin gobies do this given that they cannot see a neighboring pool? They memorize the topography of the rocks during high tide, then use their mental maps at low tide. Captive fish showed a marked improvement in orientation after an overnight opportunity to swim over the pools during an artificial high tide. Removing the gobies from their home tide-pool for various periods of time before re-testing their jumping ability showed that their memory of familiar pools lasted about 40 days. Thanks to these mental capacities, gobies caught in a shallow depression avoid having to make a pure leap of faith.

Scientists are finding that mate choice in guppies is more flexible and less 'hard-wired' than once thought. Studies by Lee Dugatkin at the University of Louisville indicate that females choose males with traits seen to be attractive to other females. It is as if a female guppy observes another female's choice and exclaims: 'I'll have what she just had.' Such behavior is more in keeping with a conscious mind than an empty one. It indicates the ability to scrutinize, compare, remember, recognize individuals, make subtle discriminations, and of course to make decisions. If these fish can be so discerning, one may wonder what they might be feeling when they choose a mate.

Fish play – not such a big leap

Is there any concrete evidence for fish pleasure? Frankly, there is practically no research into this area (yet). All one can do at this point is to recount anecdotes and appeal to circumstantial evidence. There are some good reasons to suspect that fish feel pleasure. Just as pain helps an animal that can detect and escape aversive stimuli, pleasure is useful to an animal with a sophisticated nervous system, cognitive skills like learning, long-term memory and individual recognition, and which can discern, seek out and locate important rewards.

As we saw in the discussion of fish cleaning stations in Chapter 7, some fish behave as if they are experiencing pleasure. Fishes are surely curious. They are known to cautiously investigate novel objects in their surroundings. Anyone who has had their legs nibbled at by schools of little fishes – I have on several occasions in both fresh and salt water environs – has experienced their inquisitive nature.

If fish can feel good, then perhaps the best place to look would be in the realms of play. At least one expert has predicted that, for energetic reasons, play in general might be more common in aquatic environments. *The Genesis of Animal Play* (2005) by ethologist Gordon Burghardt summarizes for the first time the scattered bits of evidence for play in fishes.

Behavior consistent with some definitions of play – including manipulating and balancing objects, leaping, and chasing games – has been reported across disparate groups of fishes. There are many accounts of fishes leapfrogging, sometimes repeatedly, over floating objects, including turtles. Play-like jumping and leaping behavior is known from at least six different families of fish. Many readers will have witnessed one or more fishes repeatedly leaping clear of the water in a lake or pond. Alternative interpretations include predator avoidance, intra-species aggression and attempts to dislodge parasites, but in the absence of more rigorous study, playful contexts cannot be ruled out.

Object play, which has been noted in captivity, is somewhat more difficult to discredit as play behavior than is the locomotor play suggested by aerial leaps. There are several compelling accounts of playful interactions with objects in captive elephantnose fish. After witnessing surprising things going on in his aquarium, the zoologist E. Meder moved it to his desk so he could keep a closer eye on his lone elephantnose fish. One day, to his astonishment, Meder watched the fish balancing a small aquatic snail on his or her nose (Meder never determined or divulged the fish's sex). Small nylon balls added to the tank were treated similarly. When these began to accumulate in the filter, Meder suspended one on a string, and the fish would bat, balance and retrieve the ball. That this fish was kept alone suggests that the play-like behavior may have been driven, in part, by boredom.

Similar observations of juggling, balancing, and other object manipulations in other species of elephantnose fish were made by the respected ethologist Monica Mayer-Holzapfel and several aquarium colleagues, in Switzerland, Denmark and Germany in the 1950s. Meyer-Holzapfel considered several non-play alternative interpretations for these behaviors, including scratching, redirected feeding, nest-building, courtship or aggression, and 'social' interactions, each of which was deemed unlikely.

Burghardt also presents some anecdotes from fishes that don't sport elephantine noses. Here are two examples. At a US zoo, captive surgeonfish (of two species) will gulp air at the surface of their tank, swim to the bottom, release the air and chase the bubbles to the surface. And the behavior of well-fed great white sharks interacting with objects hung from the sides of boats is unlike that of those who may be just after a meal.

Clearly, the evidence remains too sparse to make unequivocal conclusions that fishes play, though it does appear to occur in some species. Hopefully these examples will stimulate more biologists to take the possibility more seriously. If we take just the behavior of elephantnose fish, we may conclude, as Burghardt does, that

> The prized discontinuity between mammals and birds on the one hand and 'lower' vertebrates on the other, ... is crumbling.

Some will probably always deny that fish have feelings, including pleasurable ones. Yet, in the debate over whether they feel at all, the pendulum swings clearly in their favor. For them, as for other vertebrate animals, pleasure is adaptive.

Spineless, but senseless?

During a recent bicycle ride to work I stopped to pick up a downed spicebush swallowtail butterfly from the curbside. The insect was standing upright so I suspected she was still alive. Sure enough, she was. Experience has taught me that disabled butterflies at the roadside are often dehydrated. I dribbled a few drops from my water bottle onto my hand, and she jerked slightly as the cool water washed over her feet. Soon the butterfly's long tongue was unfurled and lapping up the water. I stood there admiring the deep velvety black and blue colors of her wings, the blue fading gradually to turquoise near the tips, and the spots of yellow and orange beneath. After three minutes of drinking she recoiled her tongue, so I placed her on a leaf and rode on.

Whenever I have close encounters with insects, I wonder if they have experiences. The single-celled amoeba needs water and moves towards it, but unlike the butterfly has no nervous system. Does it stretch credibility to think that this butterfly felt something when she flinched as the cold water touched her feet – which are studded with sense organs – or when she slaked her thirst? Could there have been any sense of relief, or pleasure?

Desiring and seeking water to relieve dehydration is expedient to any creature needing water to survive, as all animals do.

As Donald Griffin argued, we should not too hastily dismiss insects as unconscious merely because they are small and only distantly related to us. Having awareness and behaving flexibly confers great advantages over being a mechanical stimulus-responder. The latter can be harmful in a complex world. An automaton insect whose system inflexibly recognizes water as a 'flat, shiny surface' might find itself lapping at an oil slick. Finding and getting water requires fewer cognitive steps for an aware animal than for a blank one, whose system would presumably require a continuous series of yes/no loops: turn right, turn left, straight ahead, identify flat shiny surface, land, extend tongue, lap, and so on.

It is widely assumed that invertebrate animals have no feelings or emotions. It's a broad claim when one considers that some 99% of animal species are invertebrates. Of course, the weight of numbers alone need not tip the scale. There are more grains of sand on Earth than there are insects, yet I am confident that none has an ounce of feeling. However, just as the sheer weight of numbers predicts life elsewhere in the cosmos as likely, we ought to remain open to the possibility that there is feeling in an insect's feelers and sensation behind a worm's squirm.

Figure 10.1 Tiger swallowtail butterflies quench their thirst.

A recent review by British biologist Chris Sherwin from the University of Bristol throws up some challenges to traditional scientific views of the insect as a spineless, pre-programmed automaton. Sherwin found scientific evidence that invertebrates can remember and learn, have spatial awareness and mental maps, show preferences, develop habits, and respond to noxious stimuli. According to Sherwin:

> invertebrates have different sensory organs and nervous systems, and might perceive nociception or pain in an entirely different way to vertebrates, but still experience a negative mental state.

He concludes:

> the similarity of [invertebrate responses] to those of vertebrates may indicate a level of consciousness or suffering that is not normally attributed to invertebrates.

Insects can certainly learn. For example, in experiments conducted in the 1960s, when various grasshopper species were placed in the same enclosure it was expected that they would retain their songs. Instead, more than ten species imitated the songs of other types. Many insects show a plasticity of behavior that can only be explained by learning. In scientific experiments, they have been shown to avoid electric shocks, self-amputate limbs in response to apparently painful stimuli, engage in post-injury grooming, and strike at sources of pain.

Insects and some other invertebrates also show surprising parallels to vertebrates in their responses to drugs associated with pain suppression. Studies of cockroaches and crickets show that they react to a noxious stimulus (being placed on a hot plate) in the same way that vertebrates do: they run off it quickly. Give them a pain-killer (morphine) and they spend more time on the plate. Give them a drug that counteracts morphine (naloxone) and they revert to their initial quick response. Furthermore, in the cricket study, when morphine was withheld on the fifth day, the crickets showed a hyper-response, which the investigators interpreted as evidence of addiction. Morphine also produces a dose-dependent decrease in crabs' defensive response to being struck between the eye-stalks.

Opioid drugs also have an apparent analgesic effect in leeches, mollusks, crabs and insects. In many of these, naloxone reversed this analgesic effect, exactly as we would predict in vertebrates. A recent study found that crayfish respond to the rewarding properties of psychostimulants. Crayfish from a wild population in Ohio showed a significant preference for moving to a quadrant of a test arena into which was infused solutions of cocaine or amphetamine.

Insects also show taste preferences. They are equipped with cells for detecting flavors, and they are particular about what foods they eat. A study of broad-headed bugs, for example, found that they dabbed proffered foods with their antennae before probing and feeding, and their mouthparts were found to have permeable 'sensilla' which function as taste-receptors. These anatomical and behavioral assets accompany sophisticated levels of discrimination and preference. For example, a species of parasitic wasp was found to respond not only favorably to solutions of eight of fourteen naturally occurring sugars, but the insects' acceptance threshold also differed among these sugars. The other six sugars were rejected, except either when the insects were particularly thirsty, or when some were combined with trace amounts of particular sugars from the preferred group.

Exuded tree sap and rotting fruits are important feeding sources for many adult butterflies. A recent study of three tropical butterfly species found that ethanol, acetic acid and several amino acids (major components of these food sources) did not by themselves elicit feeding by foraging adults. But when mixed with the sugars that naturally accompany these comestibles, the insects began to probe and feed with their proboscises. Grasshoppers learned about food quality faster, foraged more efficiently, and consumed more food when flavors were present, suggesting that palatability increases memorability, and may prevent food-associated boredom.

Nevertheless, we should not be too hasty in assigning sensory pleasure to creatures with disparate anatomical and physiological makeup to those of us and other vertebrates. There are other studies of insects that suggest that at least some of what they do is governed by surprisingly simple neurological mechanisms. For instance, a recent study on the gustatory decisions of grasshoppers (three species) sought to corroborate a generally accepted view among insect physiologists that these decisions are based on patterns of input from populations of gustatory receptors. They conclude that the insects may be able to distinguish between different chemical compounds simply by the differential firing rate of two neurons.

Of course, our own taste sensations can be described purely mechanically in terms of inputs from populations of gustatory receptors associated with our tastebuds. And we know that this is not the whole picture of our perception of food – we know that there is an experiential side to it. The authors of the grasshopper study cited above also conclude that these two neurons 'provide qualitatively different information to the central nervous system.' Qualitative input allows the prospect of a qualitative response. Mechanistic findings in insects need not preclude some sort of sensory experience, however simple.

Why would the nectar of flowers pollinated only by insects have a sweet taste if the insects could not enjoy the taste? Certain beetles in New Guinea will stroke the bodies of mealy bug larvae for an hour or more, which return the favor by excreting sweet secretions that the beetles lap up eagerly. Some ant species protect other insects, which repay the debt by secreting sweet droplets. Examples of these nectar providers include aphids, greenflies, leaf-hoppers, and the oak blue caterpillar, which has a nipple on its back evolved for this purpose. It's a reward system, and as such, it suggests that ants experience taste pleasure. Aphids could save energy by excreting only water and keeping the sugar for themselves. Otherwise, why should the exudates be sweet? One could deny any pleasure in it by resorting to the argument that sugar water contains more energy than plain water, and that the pollinating butterfly is drawn to flower nectar, and the protective ant to honeydew, for its energetic benefits and not taste. This is not necessarily the most parsimonious explanation, for it raises the question of why the pollinators would be drawn to the plant at all.

Sensual touches?

Is it possible that some invertebrates can enjoy sex? If so, then some of them could be having a lot of fun at it. Sexual courtship is widespread and copulation is prolonged in many arthropods, including some water striders, bushcrickets, dragonflies, and dungflies, to name a few. Garden flies may remain in copulation for fifty hours or more. One millipede species copulates for an average of nearly three hours, during which time the male coils around the female and produces a large 'sexual collar' by hydraulic movements in the region of genital contact. Specific rituals involving tactile stimulation among insects and other invertebrates may play a role in averting mistaken species identity, but it also appears to aid in behavioral and physiological receptivity. The way a male flower beetle rubs the edges of a female's back during copulation, for instance, has been found to directly influence the number of sperm that fertilize her eggs.

Prolonged copulation is commonly interpreted to have evolved as a strategy to reduce competition from other males. This need not negate its being rewarding. Copulation in the soft-winged flower beetle suggests that the male may stimulate the female during mating and that successful males are more vigorous in terms of copulatory movements. A study of dungfly mating concluded that the grasping organs on the males' front legs function not to restrain females, but to stimulate them. There is a dung beetle male who caresses the female with his front pair of legs before he mounts her, then goes into spasms as he taps her back with his front legs.

Conservative views hold that these sorts of behavior are of a stimulus-response variety that lack any feeling or experience. The realms of insect existence are so removed from our own that we may never know whether or not they involve any sort of pleasure. Yet there is enough information to give pause to the idea of denying them such capacity.

Touch is important for intimate relations in a diversity of invertebrates. Female crabs become increasingly receptive to a male's gentle touch. A male fiddler crab clambers onto the back of the female he is courting and uses his large foreclaw to tap and stroke her carapace. The carapace may also be plucked in species which adorn themselves with seaweed or other materials. With his walking legs, he continuously strokes the female, on whom the effect appears hypnotic, for after a few minutes she becomes motionless. In this state, the female is flipped onto her back, from which position the two can mate, a process that lasts about one hour. Male lobsters also gently caress the female's carapace as a prelude to mating. If flies, beetles and crabs require special types of tactile stimulation to get in the mood, we may not wish to assume that the phenomenon is devoid of feeling.

The African migratory locust is infamous for its occasional population explosions, during which the normally solitary green form transforms into a highly gregarious yellow and black form. The stimulus for this pronounced and abrupt change is tactile. Physical contact with other locusts (a natural product of denser populations) causes the transformation. A series of studies at the University of Oxford found that stroking the hind leg induces the transformation from the green, solitary form to the highly gregarious, yellow and black form. Experimental stroking of other body parts failed to have this effect. The researchers were jesting when they named the sensitive area on the hindlegs the 'G-spot,' but they leave open the tantalizing possibility of good feelings.

Sexy flowers

The interplay of plants and insects offers boundless variations on relationships that have had millions of years to evolve. Many plants produce flowers that use sexual attraction to lure pollinating insects. These flowers assume the appearance of desirable mates, and manufacture 'designer drugs' that drive insects to distraction. It's a sort of spineless pornography.

The *Orphryus* orchid has evolved just the right pattern of curves, spots and hairiness to convince certain male bees that it is the abdomen of a female as viewed from behind. The effect is to induce the male to 'pseudocopulate' with the plant. The frantic copulatory movements of the insect maximize the transfer of pollen from one 'prostitute flower' to the next. Michael Pollan feels that 'bees appear lost in transports of sexual ecstasy,' compelled to 'rush around mounting one blossom after another.' Most insect-pollinated plants offer the reward of nectar in return for the insect's services. More than half of all orchid species are believed not to provide any nectar food, merely a masturbatory kick.

Not-so-humble hymenopterans

There are 10,000 ant species. Ants are so successful that their total numbers have been estimated at one quadrillion (1,000,000,000,000,000). Like bees, wasps and termites, ants are eusocial insects, so-named for their high levels of cooperation, sophisticated communication, parental care, and suicidal self-sacrifice in the colony's interests. To these tiny beasts the entomologist William Morton Wheeler in 1910 attributed 'displays of pain, anger, fear, depression, elation, and affection.' Wheeler also noted their apparent empathy when they helped crippled and distressed nestmates, and their ability to manipulate others by deception, such as by 'playing dead.'

Robert Hickling at the University of Mississippi has interpreted four distinct 'words,' or commands, made by black fire ants: (1) normal/calm (all is well), (2) alarm, (3) distress and (4) attack. The insects produce these signals by stridulation, or rubbing together specialized parts of their exoskeleton.

Such are the prodigious feats of honeybees that they have buried a number of previously held assumptions about the limitations of insect brains. Even play could be in the honeybee's behavioral repertoire. During so-called 'training flights' by recently emerged workers, individuals launch themselves from the top of the hive and flap their wings as they float to the ground; they then climb up and repeat the exercise. Honeybee authority Martin Lindauer in 1961 described the behavior as playful.

The social fighting of cockroaches of several species has also been described in the possible context of play. For example, two half-grown nymphs of the wood-feeding cockroach will square off, scuffle a few moments, then move on. Most interpretations of this behavior are agonistic, but it could be play. Cockroaches have well-developed investigatory abilities and also perform parental care.

If insects play, they may not be the only invertebrates who do. Other play-like behavior in invertebrates includes: fiddler crabs mischievously removing others' burrow lids, object play in mantis shrimps, object manipulation in lobsters, and object play in octopods (see Chapter 4).

Flying to attention

Recent studies on fruit flies indicate the rudiments of consciousness. The tiny animals' brains are only the size of a poppy seed, but that's still large enough to accommodate a quarter million neurons. When a fly is presented with a moving stimulus (a broad stripe painted on a rotating drum), her brain shows electrical activity uncannily like that of a human brain when it is stimulated. And like us, the fly begins to lose interest when the same stimulus is presented repeatedly. Change the stimulus by adding a second stripe, and the fly perks up again, even more so if its importance is enhanced, such as by simultaneously puffing the fly with banana odor. Finally, when the fly's attention is focused on the presented stimulus, she seems to ignore everything else; this sort of suppression of other distractions is the hallmark of attention, which is what neuroscientists Ralph Greenspan and Bruno van Swinderen concluded the flies' behavior was demonstrating. Other experiments from the same laboratory find that fruit flies sleep every night, they learn from positive and negative experience, and they have both short- and long-term memory.

Evidence of attention is also being documented in jumping spiders. For example, a species of Australian jumping spider has been shown to wait until the attention of its prey is diverted before trying to catch it. A series of experiments showed that *Portia fimbriata* approaches the web-spinning spider *Zosis genicularis* primarily when the latter is wrapping its own prey. Behavior like this could be pinned purely to unconscious control, but the scientists studying it think it involves thinking.

If you've never watched a jumping spider, you should treat yourself to the privilege when the opportunity arises. They are remarkably alert and behave as if they are aware. They usually re-orient themselves to face any movement when you get close. My father was completely surprised and won over when he offered a crumb to one, and the little spider came forward and took it. Two biologists who study jumping spiders – Stim Wilcox from the State University of New York, and Robert Jackson from the University of Canterbury – were moved to write:

> When we began research on *Portia*, few thoughts would have seemed more foreign to us that that one day we

would seriously be discussing cognition in a spider. Yet over and over again, *Portia* has defied the popular image of spiders as simple animals with rigid behavior.

Arthropods are not the only invertebrates causing animal behavior research to scratch its collective head. Some mollusks (which include snails, slugs, octopuses and squids) have well-developed nervous systems, and one might wonder if they can experience something like pleasure. If so, then sex may be doubly rewarding in some species, for which individuals possess both male and female reproductive organs and can thus penetrate and be penetrated at the same time. Sea-hares, large marine hermaphroditic mollusks that look like slugs, form mating chains of several individuals, the penis of one penetrating the genital pouch of the one in front. At intervals, the sea-hare at the front of the chain may break off and join the rear end of the chain. (It will be a great day for sea-hares when one of them invents the wheel.)

A study by Russian researchers found that two snail species they were studying engage in self-stimulation of the mesocerebrum, a part of the brain associated with reproductive behavior. Snails with electrodes implanted into other areas of the brain associated with withdrawal behavior generally avoided the source of the negative stimulus. The authors concluded that the snails' mesocerebrum plays 'an "emotional" role in behavior.' This finding is supported by earlier studies showing that mollusks show 'motivation' and 'arousal' to engage in feeding and sexual behavior. (Since it was first done in 1954, many researchers have studied the phenomenon of self-stimulation, in which an animal's brain tissue is penetrated by a thin electricity-conducting wire, or electrode. In the case of vertebrates a window must be cut from the skull and the animal's head immobilized in a 'stereotaxic' device. By manipulating some object in its environment, the animal receives direct electrical stimulation of the brain via the electrode. If the animal continues to manipulate the object in the presence of this experimental setup, then the stimulation is thought to be pleasant, or 'rewarding.' These unsavory experiments gave rise to the concept of 'pleasure centres.' Today, brain imaging technology, such as PET and MRI, offers a different window on the study of brain structure and function without causing grievous harm to the subjects, but the electrode method is still widely used.)

Daring to entertain ideas of cuddly crabs and sensuous snails risks opprobrium from one's scientific colleagues. Few of us, I suspect, are willing to believe that insects are conscious. And there are many examples of them behaving in an entirely mechanical way. David DeGrazia reports studies of caterpillars which climb to the tops of trees because the legs on the side of their bodies receiving more light move more slowly than the legs on the darker side. They will crawl back down and starve when their surroundings are artificially lit at ground-level and darkened overhead. Furthermore, when one eye is blinded, the caterpillars crawl in circles until they die.

It's tempting to draw a parallel with commuters getting caught in daily traffic jams of climate-killing SUVs on their way to/from work, or the proliferation of nuclear weapons, but in fairness, these are conscious, if globally maladaptive, decisions. Nevertheless, mechanical inflexibility in one type of insect need not extend to another. Bees still have mental maps, ants forage idiosyncratically, and fruit flies pay attention. Discomforting though it may be to consider that there may be some spark of feeling and experience inside an insect, some of the evidence challenges notions to the contrary.

Conclusion

Existing data suggest that fish feel, and that some invertebrates might. Behavioral studies are peeling away old assumptions and prejudices about these groups. Fish show intelligence and behavioral flexibility never before thought possible within their 'primitive' guise. And careful studies of invertebrates are finding that they, too, behave in a manner that suggests we may need to review earlier beliefs. The margins of pleasure may stretch far beyond where we once thought possible.

Part III
FROM ANIMAL PLEASURE

Chapter 11

FEELING GOOD, DOING GOOD

Implications of a pleasurable kingdom

For Nature did not idly spend
Pleasure; she ruled it should attend
On every act that doth amend
Our life's condition;
'Tis therefore not well-being's end
But its fruition

Robert Bridges (1912)

In June 2004 the BBC telecast a three-week, twelve-part nature series titled *Britain Goes Wild: Making Space for Nature*. Unlike most nature programs, this one was filmed live. One of the hosts spent the first few episodes reporting from a craggy island off the Scottish coast, where 40,000 gannets were nesting, then relocated inland to an abandoned quarry where a pair of peregrine falcons were raising their chick. Elsewhere, on an organic farm in the British midlands, remote control cameras were strategically placed allowing viewers to watch great tit and jackdaw chicks being fed inside their nest-boxes, martens making midnight raids on picnic tables, and even a fish-eye view of feathered and furred visitors to a water-bath. In a nearby wood, another cluster of cameras captured the evening routines of a family of badgers.

The badgers usually emerged at dusk, shortly before the program drew to a close. There were several young, about half the size of the adults. One might expect that after a day underground they would emerge famished, hurriedly setting off to forage. Not so. The mother badger usually began her 'working day' reclined on her rump, hind legs stretched out, having a good belly scratch. A few yards away, the young spent most of this period playing. The smallest youngster was dubbed 'Dennis' for his mischievous resemblance to the cartoon character Dennis the Menace. Dennis would perform spinning jumps and leap on the backs of his wrestling siblings. Friendly biting and tail-pulling were among his arsenal of pranks.

What struck me was the *joie de vivre* shown by these animals. Their twilight lounging and frolicking illustrate that wild nature is not an endless, joyless struggle to survive. These are not lives of quiet desperation. Wild badgers have leisure time. They take pauses. Their worlds, like ours, have stretches of tranquility, moments of pleasure, and eruptions of joy.

As it should be. We have seen that nature favors rewards and nurtures pleasure. As pain is adaptive for avoiding dangerous situations that could maim or kill, pleasure encourages behaviors that enhance survival.

Figure 11.1 Kangaroos' lives, like ours, have stretches of tranquility.

The story can end there. We can just accept that the animal kingdom encompasses a panoply of feelings from pain to pleasure, and call it a day.

Or we can parlay this awareness into some further good. The inspiration for this book arose, in part, from a dissatisfaction with the way things are. For various reasons, including ignorance, indifference, and greed, many animals get a poor deal from humans. At our best, we admire and protect them. We also eat them by the billions, cage and harm them for scientific experiments, dissect them in school classrooms, and abuse them for our recreation – in circuses, rodeos, sport fishing and hunting, to name a few. The number of feeling animals killed by humans each year is in the tens of billions. On this dizzying scale, it is easy to forget that each one is (or was) a feeling individual.

One might ask what this has to do with animal pleasure. A great deal. If animals feel, then we have responsibility towards them. And if they feel more than just pain – if they are capable of pleasure – then that responsibility is greater than if they did not.

Pleasure is for individuals

We humans have a tendency to view animals as species instead of as individuals. From this perspective it is easier to justify elephant culls, rationalize pigeon poisonings, and fill whaling quotas. In 2004, Norway announced that it was increasing its annual slaughter of minke whales from 500 to 670 kills, on the basis that recent population increases in the minke population exceeded the new quota. Viewed strictly from a population perspective, this news (if accurate) is cause for celebration. Viewed from the perspective that each minke whale is a thinking, feeling individual, it is tragic, because more animals suffer and die than before. Pleasures and pains are felt by individuals, not populations.

Viewed as a population, the domesticated chicken (*Gallus gallus*) is an unqualified success. Because of our penchant for eating them, this bird's world population far exceeds that of any other bird species. But that's not welcome news to any one of the ten billion chickens living short, wretched lives in broiler sheds and battery cages.

Thus, the danger of viewing animals only as species members is that we may lose touch with their individuality and their sensory experience of life. It harks back to the evolutionary versus the experiential contexts discussed in Part I. Strictly speaking, evolution proceeds without goal or purpose. Life in the evolutionary sense exists as a cork floating on the ocean – buffeted by wind and waves and moved by currents, but with no end goal to pursue. The same is not true of an individual's life. That life is colored by instincts, needs, emotions, desires and feelings. Unlike populations, whose existence is measured in generations, individuals experience the moment. A hippo has a good wallow. A hummingbird sips nectar from a favorite flower. A map turtle basks in the morning sun. You savor a bite of carrot cake. A baboon relaxes as a friend grooms his nape. A cow relieves an itch against a tree bole. A lovebird nuzzles with her mate.

Figure 11.2 Animals are individuals with a biography, not just a biology.

Pleasure's moral significance

Pleasure is central to human values. Most of our time and energy is spent pursuing it, at least in more affluent societies where we can afford to. We work to earn money so that we can attain or maintain a rewarding lifestyle. We play games and sports, watch television, throw parties and dine out, all chiefly for pleasure's sake. Success tends to be measured by material wealth. Of course, it isn't all about sensory pleasure. Many of us strive to do what's right, to leave a legacy, to fulfill lifetime goals such as having children (or not), developing a business or saving a habitat. It's another case of feeling good – this time by doing good.

Pleasure is also central to human morality. The reason that so many crimes – theft, vandalism, blackmail, murder – are wrong is that, ultimately, they deprive the victims and their loved ones of the opportunity to live their lives fully, to enjoy life. Consider the following hypothetical conversation between a human and a curious alien (or if you prefer, a naïve young child):

Alien/Child: Why is it wrong to kill another person?
Human: Murder is wrong because it deprives the other person of his or her life.
Alien/Child: Why is that a bad thing?
Human: It is bad because people want to live.
Alien/Child: And why so?
Human: Because life is worth living, because we can enjoy life. Life presents many opportunities to experience good things, and it is wrong to deprive others of them.

A primary aim of this book has been to show that life for many other animals is also worth living. Deny an animal basic needs (e.g. food, social contact, shelter) and we deny it basic pleasures. Studies on pigs' social behavior conducted at Purdue University have found that they crave affection and are easily depressed if isolated or denied playtime with each other. This is reduced welfare by denial of pleasure.

Without openly acknowledging it, we are already looking out for animals' pleasure. In many parts of the world, laws are being passed which not only recognize animals' capacity for pain and suffering, but their capacities for pleasure and happiness as well. In Europe, EU member nations must phase out veal crates and battery cages for laying hens. By 2012 the keeping of pregnant pigs in stalls that do not allow them room to turn around must give way to open-air stalls. England's train from York to London passes such a piggery, where I've seen litters of piglets running on their plots of grass and sows lying out nearby. It's a far sight better than being chained in a dark, stinking concrete warehouse. A 2001 European Commission Directive also requires that pig farmers furnish their animals with 'manipulable materials' like straw, wood and sawdust, the aim being to satisfy pigs' natural urge to root.

EU laws set minimum requirements; they don't prevent individual states from taking further steps. Germany plans to phase out mass farming of caged chickens by the end of 2006. In 1998 the Swedish parliament enacted a law to entirely phase out intensive confinement (factory) farms.

These reforms don't just apply to animals raised for meat. A sweeping animal protection law passed in Austria in 2004 ensures that puppies and kittens no longer swelter in pet shop windows, outlaws the use of lions and other wild animals in circuses, and makes it illegal to restrain dogs with chains, choke collars or 'invisible fences' that administer electric shocks. In March 2004, Hungary's parliament banned cockfighting and the breeding or sale of animals for fighting, and upgraded animal torture from a misdemeanor to a felony punishable by up to two years in prison. In 2003 the region of Catalonia, which passed Spain's first animal cruelty law in 1988, raised animal cruelty fines to as much as $2,400. Sweden banned ear cropping and tail docking of dogs in 1989. Italian law changed in 1994 to support a student's right to object to harmful animal experimentation; in 2005, the city of Rome joined Turin in legally requiring daily walks for dogs, and banned keeping fish in spherical fishbowls.

As these examples show, most animal welfare legislation aims to reduce pain and suffering, not maximize pleasure. Yet, to the extent that pain and pleasure lie on a continuum, they may grant animals more opportunities to feel good.

In the USA, the pattern is similar. In 2002 the US government passed the Chimpanzee Health Improvement, Maintenance and Protection (CHIMP) Act, which establishes a retirement home for some of the approximately 1,500 chimpanzees in laboratory experimentation there. In November 2002, Florida voters also supported a ban on sow confinement units, marking the first time in US history that a factory farming practice has been outlawed. In mid-2005, Oregon became the eleventh state to uphold a student's right to learn biology without harming or killing animals. Amendments to the United States Animal Welfare Act encourage animal research facilities to provide 'environmental enrichments' (e.g. toys, climbing objects, puzzles) for caged primates. Another amendment stipulates providing daily exercise for dogs, on the understanding that they enjoy being taken for walks. These are fairly minimal gains for animals who spend their lives in small cages or rooms and are used for experiments rarely carried out in their best interests. They do, however, recognize a quality of life.

In one case, even invertebrate feelings are now given legal consideration. The United Kingdom's Animals (Scientific Procedures) Act, originally passed in 1986, includes in its purview a mollusk: the common octopus (*Octopus vulgaris*).

Moral animals – fairer is fitter

Clearly, some aspects of our animal welfare laws reflect a concern for the role of pleasure in their quality of life. What about how animals treat each other? Given pleasure's important role in adaptation and survival, we might expect natural selection to favor consideration for others. And so it does. Other species – especially social ones – do have codes of conduct. Successful group living requires getting along with your group-mates. Individuals who don't behave accordingly may be shunned by the group. Primates of many species appear to have a sense of right and wrong, fair and unfair. At least one chimp expert believes that chimps are as socially sophisticated as humans. Frans de Waal has also suggested that 'reciprocity among chimps is governed by the same sense of moral rightness and justice as it is among humans.' Shirley Strum described the olive baboons she studied as 'social diplomats.' A recent study with captive capuchin monkeys found that they are more likely to reject a cucumber slice after seeing that another capuchin has received a more attractive grape. These observations indicate that the monkeys recognize inequality, and are willing to give up some material pay-off in hopes of getting a more equitable outcome: being handed a grape by the investigator who just gave one to the other monkey.

Marc Bekoff, an ethologist and leading thinker in the study of animal morality, believes morality evolved because it is adaptive:

> Nobody has really considered the possibility that being considerate to your neighbors [i.e. someone not genetically related] might sometimes be the best way to survive. But I'm starting to find evidence that a well-developed sense of fair play helps non-human animals live longer, more successful lives.

Bekoff suggests that play behavior may be a 'foundation of fairness.' Such a foundation might provide the rudiments of a sense of right and wrong in animals. Consider the social consequences we humans face if we don't play fairly? We risk losing playmates, partners, friends, and, depending on the severity of the transgression, fines and imprisonment. Successful play requires fair play. An animal who bites too hard during a play-fight may find herself engaged in a real fight. A larger individual who plays too roughly with a smaller one may find herself without a playmate. This is probably why several species are known to adjust the intensity of their play-fighting to maintain the play mood.

In other words, virtue is it's own reward. Fairer is fitter.

Fair play requires paying heed to – or at least permitting – the expression of not only one's own pleasure, but that of another. It means that if you do something nice or helpful for me, I'm more inclined to return the favor. You scratch my back and I'll scratch yours. It is a sort of non-binding social contract.

Animals not only act nice towards others; they may also go to great lengths to avoid causing them harm. A series of horrible psychology experiments conducted in the 1960s on rhesus monkeys found that they would sacrifice their own interests to protect those of another. Some monkeys went without food for days rather than cause pain, in the form of an electric shock, to another monkey. Monkeys who had previously experienced shock were more likely to sacrifice food rewards, indicating the influence of empathy on their decisions. Rats have also been shown to restrain themselves when they know their actions would cause pain to another individual.

There are countless anecdotal examples of animals showing consideration for others. Dave Sidden, owner of a wildlife rehabilitation center in Oregon, recalls his horror when he saw the smallest of a litter of four orphaned kittens totter into the neighboring enclosure of an orphaned 560-pound grizzly bear. 'At the time, "Griz" was eating lunch. When Griz saw "Cat" approaching, he removed a small piece of chicken from his bowl and placed it on the floor, where Cat promptly ate it. After that, the two orphans became close friends, eating, sleeping and playing together.'

When the kitchen of a dog kennel in Battersea, England, began getting regularly ransacked, the staff erected a hidden camera to film the goings on at night. They soon discovered that a lurcher dog named Red was using his mouth to unlatch his kennel lock and gain access to the kitchen's delicacies. But Red wasn't making a bee-line for the grub; he first opened the locks of his fellow mutts' cells so they could join the party.

A young sparrow collided with a window at a zoo, and a chimp quickly retrieved the stunned bird and seemed delighted to hold the little creature as other chimps gathered round. The fledgling was passed around hand to hand, and eventually handed through the bars to the keeper who had been watching.

In experiments in which two chimps must cooperate to access food, they do so. In one such experiment at the Language Research Center of Georgia State University, a chimp named Sherman had access to several locked boxes containing food. Austin, in an adjoining room connected by a small window, had tools to open the boxes. That Sherman and Austin learned to cooperate to gain access to the food was not surprising. Chimp societies are renowned for the formation of alliances and reciprocity, so such behavior comes naturally to them. More revealing was that after the correct tool was passed to Austin, he opened the container and passed food to Sherman. This controlled experiment illustrates consideration for the wants and needs of another.

Perhaps Sherman is improving his relations with Austin, which could be useful at some future time when the tables are turned and he needs Austin's help. This sort of 'reciprocal altruism' has become an important area of study in animal behavior since it was first named in 1971. A simpler, more practical explanation is that Sherman and Austin are friends, and friends share. Neither theory excludes the other. Nor do they preclude the likelihood that there is pleasure in sharing. Giving feels good. Certainly the brain's pleasure centers are strongly activated when people cooperate.

Scientists are now examining moral behavior in animals as never before. Primatologist Frans de Waal has given us *Good Natured: The Origins of Right and Wrong in Humans and Other Animals*. And Bekoff has written *Wild Justice, Cooperation, and Fair Play: Minding Manners, Being Nice, and Feeling Good*. These are refreshing developments. Perhaps the day will come when a criminal or madman is no longer described as 'behaving like an animal.'

Lives that matter

The behaviors, sensory abilities, and flexible lives of many of our animal kin suggest beings who are not merely alive but who have a life. As American philosopher Tom Regan puts it, they have not merely a biology but a biography. They are lives experienced across the pain–pleasure continuum – lives made better or worse by their circumstances. As Burger's relationship with Tiko matured, she came to feel that 'his life was as important as mine, his desires and inclinations equally valid.' Burger became convinced by witnessing the richness of Tiko's life – his likes and dislikes, his moods, his fallibility.

If a creature can feel pain and/or pleasure, then it has interests. It has an interest in avoiding painful situations and, where possible, in maintaining pleasurable or at least comfortable ones. These sorts of interests don't hinge on moral awareness. Human infants are not morally aware; yet they are sentient, and they deserve moral consideration. It makes no difference if the infant is a future princess or a penniless orphan; either way, the little being is entitled to some protection. Similarly for a rabbit, a chicken or a goldfish: sentience demands moral consideration.

If we believed that animals could only feel pain and suffering, then in the absence of pain and suffering all would be well. If we recognize animals' capacity for pleasure, then we may conclude that it's wrong to deprive them of it.

Plutarch recognized this nearly two thousand years ago when he reflected on the killing of animals for food:

> But for some little mouthful of flesh we deprive a soul of the sun and light, and of that proportion of life and time it had been born into the world to enjoy.

Ruth Harrison recognized it two thousand years later in her book on factory farming:

> Have we the right to rob them of all pleasure in life?

By questioning our right to deprive animals of pleasure – as we do when we confine them in tiny cages (think laboratory), torment them (think bullfight) or kill them (think lamb) – I am not suggesting that we have a responsibility to actively ensure that they attain it. Surely we have no moral responsibility to interfere with free-living animals to ensure they attain pleasure, except perhaps when our own activities deprive them of the opportunity to lead natural lives, such as when we destroy or pollute their habitats. But we do, it seems, have a responsibility not to frivolously deprive other feeling animals of opportunities to reap the rewards that life offers.

It helps to distinguish negative from positive rights here. Negative rights are those rights we shouldn't deny animals, such as life, freedom, and the opportunity to seek a fulfilling natural life. Asserting the negative rights of animals may be a passive process; for instance, we uphold such rights when we choose not to interfere with wild animals. Or they may be more active, such as when we try to correct those harms we may already have done; rejuvenating a formerly drained wetland habitat would be an example.

In the case of animals already subjugated to our service, we may also consider the assertion of 'positive rights.' These are rights which moral conduct demands we grant animals in the interests of making their situation better. In the case of companion animals, we may take steps to try to increase their opportunities for a more enjoyable life, by providing them with toys, regular exercise, food treats and so on. Similarly, for farmed animals we may seek to give them opportunities to engage in rewarding activities by providing them with more space, play objects, companionship etc. The industry term for this is 'environmental enrichment.' Putting a nest box in the mouse cages of a laboratory constitutes environmental enrichment. Mice show a strong preference for cages with a nestbox over those without, and the shelter affords them the expression of highly motivated behaviors – in this case hiding and building nests (which both male and female mice do). Of course, it's still a far cry from the ideal living environment, which would allow them to forage, climb, dig, explore, run, and choose compatible social groups, and which typical laboratory housing today routinely denies animals.

With rights come responsibilities, don't they? Actually not. The philosophy of rights distinguishes two categories of individuals worthy of moral consideration: moral agents and moral patients. Moral agents are individuals with an awareness of right and wrong. They are capable of making morally informed decisions. Humans certainly qualify as moral agents, as perhaps do some other primate species. Moral patients, on the other hand, are individuals without moral awareness but who nevertheless may be deemed worthy of moral consideration, based usually on their sentience, and in turn, their eligibility for quality of life. Most vertebrate animals also meet these criteria, though our laws and practices do not yet reflect this. To the extent that we can gauge an animal's capacity for moral awareness, most vertebrate species may be classified as moral patients. Our moral and legal codes place mentally enfeebled human beings and human infants in this category, but, as yet, no animals.

In the end, though, the presence or absence of moral behavior has little bearing on moral worth. At its worst, human behavior is among the least moral – witness colonialism, slavery, war, ethnic cleansing, factory farming, whaling, terrorism and the extinction of species. Yet immoral behavior does not denote an absence of moral awareness, and humankind's moral lapses by no means deny us moral consideration. What gives us moral standing is not so much that we can reason as that we can feel. And so it goes – or should go – for other animals.

Hedonic ethology

One of this book's aims is to inspire more interest in the study of positive experience in animals. There are great and growing numbers of accounts and scientific studies that support animal pleasure in a broad spectrum of species. Such is the wealth of evidence that few scientists would deny outright that animals experience pleasure. Yet many, if not most, continue to avoid the subject in their day jobs, be it in the papers they publish in scientific journals, or the lectures they give to students or colleagues.

In Part I, I took science to task for focusing only on evolutionary interpretations of animal behavior and neglecting their experiences. Redressing this imbalance would, I believe, benefit our understanding of animal behavior and its evolution. The capacity for pleasure informs interpretations of animal behavior in conscious, feeling individuals.

Pleasure makes an interesting focus for the study of animal behavior. The study of animal pleasure – it might be termed *hedonic ethology* – is the behavioral study of positive experiences in animals. (The adjective 'hedonic' derives from the Greek *hedone*, meaning pleasure; 'hedonism' is a more familiar variant.) Hedonic ethology acknowledges that animals are aware, and that they are sentient. It uses the same sorts of methods that ethologists use, but its focus is to examine evidence for rewarding experiences in its subjects.

Hedonic ethology ought to be an ethical science that regards the animals' interests as paramount. To deem otherwise would undermine its premise, which is that animals feel and that their lives matter to them. Deliberately harming animals to try to improve our understanding of their capacity for pleasure seems self-defeating and unethical. Some of the scientific experiments I have drawn from in this book to support my case for animal pleasure have involved procedures harmful or otherwise inhumane to the animal subjects used. By citing these studies, it is not my intention to endorse such methods, or to encourage future studies that harm animals. To the extent we recognize other animals as being capable of feeling pleasure and joy, and pain and suffering, we also assume some moral responsibility to treat them accordingly. It is sentience – not language, architecture, or a proficiency with chess openings – that crucially qualifies an individual for moral protection. Sentience, and pleasure in particular, are accessible to science without causing animals pain and suffering. We can, and ought to advance our knowledge of animal pleasure, and we should do so in the same spirit that we conduct scientific studies on human subjects – always with their interests in mind.

An organization already exists whose aim is to promote the ethical study of animal behavior. In 2000, Marc Bekoff joined forces with Jane Goodall to form Ethologists for the Ethical

Treatment of Animals/Citizens for Responsible Animal Behavior Studies (EETA/CRABS). The organization's mission is:

> ...to develop and to maintain the highest of ethical standards in comparative ethological research that is conducted in the field and in the laboratory. Furthermore, we wish to use the latest developments from research in cognitive ethology and on animal sentience to inform discussion and debate about the practical implications of available data and for the ongoing development of policy.

Like EETA/CRABS, hedonic ethology would build on a growing awareness of animal sentiency, and develop the idea that animals are pleasure-seekers into a more ethical science of animal behavior. Ethology is growing in prominence as a scientific discipline. Its tradition of studying the whole animal (not bits of animals) and striving to understand their behavior in the natural setting has the potential to advance our appreciation for animal pleasure like no other field.

Science now has the technology to probe secrets of animals' lives that were unavailable before, using methods that cause minimal disturbance. We now have subtle, non-invasive means of identifying and monitoring the activity of free-living individuals. We can hide small cameras in strategic places – including mobile ones, like the rolling 'dung-cams' and 'mirror-cams' used to study elephants and bears, respectively; we can use satellites to trace the global movements of ocean wanderers; and using just a sample of fur or a feather we can determine the genetic relatedness between individuals. Researchers are boundlessly creative once they put their minds to a problem, and I have no doubt that many ingenious approaches are in store for the nascent study of animal pleasure.

Here are two examples of current studies that might fall under the mantle of hedonic ethology.

Five heifers, aged 7–12 months, were trained to press a panel to open a gate for access to a food reward. At critical points on the animals' learning curves, behavioral signs of excitement (jumping, bucking, or kicking) were recorded (by hidden cameras), and the animals' heart rates rose (as recorded by non-invasive monitoring equipment). In a control group of five heifers whose access to food was provided independently of their panel presses, no play-like behaviors were recorded. This study suggests that cows can have 'eureka' moments, taking pleasure in their own learning achievements.

Hens were tested with colored buttons. If they pecked on one of the buttons, they received a food reward three seconds later. If they took the reward immediately, then they would receive no more, but if they waited for 22 seconds, they were rewarded with a food 'jackpot.' The hens held out for the jackpot 90% of the time. The investigators concluded that chickens live not just in the present, but that they anticipate future events and can exercise self-control to optimize rewards.

These studies involved tame, domesticated animals. Because of their tractability, domesticated animals make good subjects for studying pleasurable experiences. They are usually more accustomed to the proximity of familiar humans, although minimizing the human presence to avoid 'observer effects' on the animals is best.

While more challenging, the study of wild animals in natural settings has the important advantage of lessening or eliminating the intrusions of artificiality. Behaviors of undisturbed animals in their natural environments can more reliably be attributed to the animal's true biology and not some artifact of captivity. As biologists Marc Bekoff and Colin Allen state in the *Encyclopedia of Animal Behavior*:

> The importance of studying animals under field conditions cannot be emphasized too strongly. Field research that includes careful and well-thought-out observation,

> description, and ethically sound experimentation that does not result in mistreatment ... is extremely difficult to duplicate in captivity.

It follows that hedonic ethology should also try to steer clear of capturing wild animals and studying them in captivity. Animals kept in impoverished and/or confined environments commonly develop abnormal behaviors called 'stereotypies.' These repetitive, unvarying and apparently functionless behavior patterns arise from the frustration of natural behaviors that the animals are highly motivated to perform. If you've seen a big cat or a bear repeatedly pacing back and forth at the zoo, you've seen a stereotypy. Stereotypies – including repeatedly running back and forth in a repetitive pattern, head swaying, and mouthing or gnawing cage wires or bars – are also widespread in animals confined in factory farms, laboratories, circuses, and fur farms, and have been linked to stunted brain development, impaired learning and suffering.

Nevertheless, as long as wild animals *are* being kept in captive situations, as they are in zoos and aquaria, judicious study of pleasurable experience may be beneficial to them. Boredom is one of the most serious welfare problems for captive animals (recall, for instance, Pigface the soft-shelled turtle in Chapter 4). To the extent that captive studies can provide positive stimulation (e.g. foraging challenges, social engagement, and opportunities for play) for animals faced with the monotony and confinement of captivity, a zoo-based branch of hedonic ethology may be encouraged.

Hedonic ethology can also be thought of as an offshoot of the fast-growing field of animal welfare science. One of its most influential thinkers is Marian Dawkins, who introduced the idea of preference testing and the application of economic demand theory to animal behavior. In preference testing, animals are given a choice over certain aspects of their environment. This is usually done in captive situations, where scientists can exercise more control over their investigations, but certain preferences might also be studied in free-living animals, either by experimental manipulations (e.g. by offering food choices, choices of different nesting materials, different sounds, smells, colors etc.) or by close observations of natural choices.

Economic demand theory provides a somewhat more nuanced assessment of what animals want, by requiring them to perform some task to gain access to one or more resources. To the degree that a hen is willing to push her way through a weighted doorway to gain access to a dustbath, or that a rat is willing to work much harder (pressing a lever over 70 times) to gain access to other rats than to gain access to a larger cage or a cage with novel objects, we may conclude that these are highly desirable resources. A review of 40 studies published between 1987 and 2000 concluded, not surprisingly, that mice prefer more complex cages, and will work for nesting material, shelter, raised platforms, a running wheel and larger cages. Mice spent significantly less time in a cage with 'enrichments' (toys) than in a cage with food, extra space and shelter, when access to each cage carried the cost of having to traverse varying lengths of 2 cm deep water.

The great innovation of these approaches is that they emphasize the animals' point of view. However, there are caveats. One of the challenges of interpreting preference tests in captive animals is that they may be choosing the lesser of two undesirable options. Studies of free-living animals are less vulnerable to this problem because they are free to shun all presented options.

Finally, I hope I've illustrated amply the power of anecdote to inform us about animals' private pleasures. One of my favorites comes from naturalist-photographer Lewis Wayne Walker, who discovered a wild rat running in a rodent exercise wheel he had stored in his barn. By itself, it's just an isolated, if compelling, note. But what's to keep us from setting out running wheels (instead of traps) in places where rats live and monitoring the results?

Towards a pleasurable kingdom

Until recently, prevailing scientific dogma rejected the notion that animals have minds and feelings. The window on the question of animal pleasure was closed and shuttered. The shutters are now open. While we may never be able to communicate with other animals as clearly as we can communicate with other humans, their behavior is a window through which we can learn much about their private feelings. From the play of animals to their food preferences, from their sexual behavior to their anticipation of rewards, it is clear that they, too, have richly positive experiences. We are not the only species to feel exhilaration's rush, anticipation's pull, or paroxysms of delight.

With the shutters removed, it is now time to open the windows and let the air in. We need a science of animal pleasure that perceives animals in a fresh, new way. We need what biologist E. O. Wilson refers to as 'biophilia.' Biophilia literally translates to 'love of life.' It captures our natural inclination to relate to and appreciate nature. It is what draws us to the beach, stirs us to walk in the woods, to picnic on a hillside, and to gaze at sunsets or at insects.

In 1960, Adriaan Kortlandt made the following observation from an 80-foot-high tree platform, during an expedition to study chimpanzees on a plantation in the Belgian Congo:

> I saw a chimpanzee gaze at a particularly beautiful sunset for a full 15 minutes, watching the changing colors until it became so dark that he had to retire to the forest without stopping to pick a papaw for his evening meal.

The poignancy of this report is that it reminds us of ourselves. The chimp became so preoccupied that he just couldn't pull himself away until it became too late and he was forced to rush. He may have been aware that he needed to allow time to collect a papaw, or perhaps he just became lost in thought, forgetting altogether about the fruit – distracted from a thing of value by a thing of beauty.

His story is also a tale of animal pleasure. The gazing chimp was a creature relaxed, comfortable, and contented with life. Not in a hurry, not hungry, not fleeing any danger. He had time to rest and meditate, to put aside his cares and responsibilities.

That chimpanzee is an emblem for a pleasurable kingdom. The animal world is teeming with an enormous variety of breathing, sensing, feeling creatures who are not merely alive, but living life. Each is trying to get along – to feed and shelter themselves, to reproduce, to seek what is good and avoid what is bad. There's a diversity of good things to be gotten: food, water, movement, rest, shelter, sunshine, shade, discovery, anticipation, social interaction, play and sex. And because gaining these goods is adaptive, evolution has equipped animals with the capacity to experience their rewards. Like us, they are pleasure-seekers.

BIBLIOGRAPHY

The following sources are provided in the order mentioned, by chapter. Where a source is cited more than once in a chapter, only the first citing is shown.

Chapter 1

Wood Krutch, J. (1956) *The Great Chain of Life*. Boston: Houghton Mifflin.

Billings, J. and Sherman, P. W. (1998) Antimicrobial functions of spices: why some like it hot. *Quarterly Review of Biology*, **73**, 3–49.

Heinrich, B. (1999) *Mind of the Raven: Investigations and Adventures with Wolf-Birds*. New York: HarperCollins, p. 286.

Cabanac, M. (1971) Physiological role of pleasure. *Science*, **173**, 1103–1107.

Bekoff, M. (2000) Iguanas register pleasure. *New Scientist*, 29 April, p. 32.

Von Muggenthaler, E. (2001) The felid purr: a bio-mechanical healing mechanism. Paper presented at the 142nd annual Acoustical Society of America, American Institute of Physics, International Conference. [cited Oct 2002] Available from: http://www.animalvoice.com/catpur.htm.

Leung, K. S., Lee, W. S., Tsui, H. F., Liu, P.P. and Cheung, W. H. (2004) Complex tibial fracture outcomes following treatment with low-intensity pulsed ultrasound. *Ultrasound in Medicine and Biology*, **30**, 389–95.

Harrison, D. (2001) Feline purring shown to be effective vibrational energy healer. *The Daily Telegraph*, 22 March. [cited April 2003] Available from: http://www.telegraph.co.uk/connected/main.jhtml?xml=/connected/2001/03/22/ecncat22.xml.

Romanes, G. J. (1884; reprinted 1969) *Mental Evolution in Animals*. New York: AMS Press.

Huxley, J. (1938) *Animal Language*. London: Country Life.

Hartshorne, C. (1973) *Born to Song: An Interpretation and World Survey of Bird Song*. Bloomington: Indiana University Press.

Skutch, A. (2000) Singing the praises of family. In: Bekoff, M. (ed.) *The Smile of a Dolphin: Remarkable Accounts of Animal Emotions*. New York: Discovery Books, pp. 52–3.

Pani, A. K. and Anctil, M. (1994) Evidence for biosynthesis and catabolism of monoamines in the sea pansy *Renilla koellikeri* (Cnidaria). *Neurochemistry International*, **25**, 465–74.

Lett, B. T. and Grant, V. L. (1989) The hedonic effects of amphetamine and pentobarbital in goldfish. *Pharmacology Biochemistry & Behavior*, **32**, 355–6.

Pani, L. and Gessa, G. L. (1997) Evolution of the dopaminergic system and its relationships with the psychopathology of pleasure. *International Journal of Clinical Pharmacology Research*, **17**, 55–8.

Panksepp, J. (1998) *Affective Neuroscience*. Oxford: Oxford University Press.

Berridge, K. (1996) Food reward: brain substrates of wanting and liking. *Neuroscience and Biobehavioral Reviews*, **20**, 1–25.

Siviy, S. M. (1998) Neurobiological substrates of play behavior: glimpses into the structure and function of mammalian playfulness. In: Bekoff, M. and Byers, J. (eds.) *Animal Play: Evolutionary, Comparative, and Ecological Perspectives*. New York: Cambridge University Press.

Bekoff, M. (2002) *Minding Animals: Awareness, Emotions, and Heart*. Oxford: Oxford University Press.

Bekoff, M. (ed.) (2004) *Encyclopedia of Animal Behavior*. Westport, CT: Greenwood Press.

Goodall, J. (1986) *The Chimpanzees of Gombe: Patterns of Behavior*. Cambridge, MA: Belknap Press.

de Waal, F. (1996) *Good Natured: The Origins of Right and Wrong in Humans and Other Animals*. Cambridge, MA: Harvard University Press.

Payne, K. (1989) Elephant talk. *National Geographic*, August, pp. 264–277.

Kruuk, H. (1972) *The Spotted Hyena: a Study of Predation and Social Behavior*. Chicago: University of Chicago Press.

Ford, J. K. B. (1984) Call traditions and dialects of killer whales (*Orcinus orca*) in British Columbia. *Ph.D. Thesis*. University of British Columbia.

Hauser, M. (2001) *Wild Minds: What Animals Really Think*. London: Penguin, p. 195.

Knutson, B., Burgdorf, J. and Panksepp, J. (1998) High-frequency ultrasonic vocalizations index conditioned pharmacological reward in rats. *Physiology & Behavior*, **66**(4), 639–43.

Mason, G., Cooper, J. and Clarebrough, C. (2001) Frustrations of fur-farmed mink. *Nature*, **410**, 35–6.

Dawkins, M. S. (1998) Evolution and animal welfare. *Quarterly Review of Biology*, **73**, 305–28.

Chesler, E. J., Wilson, S.G., Lariviere, W. R., Rodriguez-Zas, S. L. and Mogil, J. S. (2002) Identification and ranking of genetic and laboratory environment factors influencing a behavioral trait, thermal nociception, via computational analysis of a large data archive. *Neuroscience and Biobehavioral Reviews*, **26**, 907–23.

Willenbring, S. and Stevens, C. W. (1995) Thermal, mechanical and chemical peripheral sensation in amphibians: Opioid and adrenergic effects. *Life Sciences*, **58**, 125–33.

Sneddon, L. U., Braithwaite, V. A. and Gentle, M. J. (2003) Do fishes have nociceptors? Evidence for the evolution of a vertebrate sensory system. *Proceedings of the Royal Society of London B*, **270**, 1115–21.

Morris, D. (1990) *Animalwatching: A Field Guide to Animal Behavior.* London: Jonathan Cape.

Persinger, M. A. (2003) Rats' preferences for an analgesic compared to water: an alternative to 'killing the rat so it does not suffer.' *Perceptual and Motor Skills*, **96**, 674–80.

Colpaert, F. C., De Witte, P., Mondi, A. N., Awouters, F., Niemegeers, C. J. E. and Jansen, P. A. (1980) Self-administration of the analgesic suprofen in arthritic rats: evidence of *Mycobacterium butyricum*-induced arthritis as an experimental model of chronic pain. *Life Sciences*, **27**, 921–8.

Colpaert, F. C., Meert, Th., De Witte, P. and Schmitt, P. (1982) Further evidence validating adjuvant arthritis as an experimental model of chronic pain in the rats. *Life Sciences*, **31**, 67–75.

Beukema, J. J. (1970a) Angling experiments with carp (*Cyprinus carpio* L.). II. Decreased catchability through one trial learning. *Netherlands Journal of Zoology*, **19**, 81–92.

Beukema, J. J. (1970b) Acquired hook avoidance in the pike *Esox lucius* L. fished with artificial and natural baits. *Journal of Fish Biology*, **2**, 155–60.

Ehrensing, R. H., Michell, G. F. and Kastin, A. J. (1982) Similar antagonism of morphine analgesia by MIF-1 and naxolone in *Carassius auratus*. *Pharmacology Biochemistry & Behavior*, **17**, 757–61.

Danbury, T. C., Weeks, C. A., Chambers, J. P. and Waterman-Pearson, A. E. and Kestin, S. C. (2000) Self-selection of the analgesic drug, carprofen, by lame broiler chickens. *Veterinary Record*, **146**, 307–11.

Sapontzis, S. (2002) Unethical considerations: Probing animal research. *AV Magazine*, **110**(2), 6–9.
Burns, R. (1786) *Poems, Chiefly in the Scottish Dialect*. Kilmarnock: John Wilson.

Chapter 2

Linden, E. (2003) *The Octopus and the Orangutan: New Tales of Animal Intrigue, Intelligence, and Ingenuity*. London: Plume.
Berrill, N. J. (1969) quoted in Krutch, J. W. *The Best Nature Writing of Joseph Wood Krutch*. New York: William Morrow & Co., p. 176.
Voilquin, J. (1965) French translation of *Ethique de Nicomaque* 10. Paris: Flammarion.
Descartes, René (1901) *The Discourse on Method and Metaphysical Meditations*, transl. G. B. Rawlings. London: Walter Scott.
Darwin, C. (1859) *On the Origin of Species*. London: John Murray.
Darwin, C. (1998) *The Expression of the Emotions in Man and Animals*, 3rd edn. (ed. Ekman, P.). London: HarperCollins.
Larson, E. J. (1998) *Summer for the Gods: The Scopes Trial and America's Continuing Debate over Science and Religion*. Cambridge, MA: Harvard University Press.
Budiansky, S. (1998) *If a Lion could Talk: Animal Intelligence and the Evolution of Consciousness*. New York: The Free Press.
Damasio, A. (2000) *The Feeling of What Happens: Body, Emotion and the Making of Consciousness*. London: Vintage.
Griffin, D. R. (1991) Progress toward a cognitive ethology. in Ristau, C. A. (ed.) *Cognitive Ethology: The Minds of Other Animals*. Hillsdale, NJ: Lawrence Erlbaum Associates, pp. 3–17.
Macphail, E. (1998) *The Evolution of Consciousness*. Oxford: Oxford University Press, p. 136.
Dennett, D. C. (1996) *Kinds of Minds: Towards an Understanding of Consciousness*. London: Phoenix.
Panksepp, J. (2005) Toward a science of ultimate concern. *Consciousness and Cognition*, **14**, 22–9.
Louie, K. and Wilson, M. A. (2001) Temporally structured replay of awake hipppocampus ensemble activity during rapid-eye movement sleep. *Neuron*, **29**, 145–56.
Griffin, D. R. (1976) *The Question of Animal Awareness*. New York: Rockefeller University Press.
Griffin, D. R. (1984) *Animal Thinking*. Cambridge, MA: Harvard University Press.

Griffin, D. R. (1992) *Animal Minds*. Chicago: University of Chicago Press.
Fox, D. (2004) Do fruit flies dream of electric bananas? *New Scientist* **181**, 32–5.
Bekoff, M. and Sherman, P. W. (2004) Reflections on animal selves. *Trends in Ecology and Evolution*. **19**, 176–80.
Bullock, T. H. (1977) *Introduction to Nervous Systems*. San Francisco: W. H. Freeman.
McMillan, F. D. (2004) Psychological well-being. In: Bekoff, M. (ed.) *Encyclopedia of Animal Behavior.* Westport, CT: Greenwood Press, p. 1134.
Maddock, A. (1971) *Animals at Peace*. New York: Harper & Row.
Dawkins, R. (1995) *River out of Eden: A Darwinian View of Life*. London: Phoenix, p. 154.
Cartmill, M. (1996) *A View to a Death in the Morning: Hunting and Nature Through History*. Cambridge, MA: Harvard University Press.
Wright, R. (1994) *The Moral Animal: Why We Are the Way We Are.* London: Abacus, p. 163.
Cone, M. (2003) Of polar bears and pollution. *New York Times*, 19 June.
Howard, W. E. (1990) *Animal Rights vs. Nature*. University of California: Walter E. Howard.
Tannenbaum, J. (2001) The paradigm shift toward animal happiness: what it is, why it is happening, and what it portends for medical research. In: Paul, E. F. and Paul, J. (eds.) (2001) *Why Animal Experimentation Matters: The Use of Animals in Medical Research.* New Brunswick: Transaction Publishers, pp. 93–130.
Burghardt, G. M. (1991) Cognitive ethology and critical anthropomorphism: a snake with two heads and hognose snakes that play dead. In: Ristau, C. A. (ed.) *Cognitive Ethology: the Minds of Other Animals*. San Francisco: Erlbaum, pp. 53–90.
Bekoff, M. (2004) Wild justice, cooperation, and fair play: minding manners, being nice, and feeling good. In Sussman, R. and Chapman, A. (eds.) *The Origins and Nature of Sociality.* Chicago: Aldine, pp. 53–79.
de Waal F. (2001) *The Ape and the Sushi Master.* New York: Basic Books.
Black, J. M. (ed.) (1996) *Partnerships in Birds: The Study of Monogamy.* Oxford: Oxford University Press.
Howard, C. J. (1995) *Dolphin Chronicles*. New York: Bantam.

LeDoux, J. (1996) *The Emotional Brain*. New York: Simon & Schuster.
Lorenz, K. (1963) *Studies in Animal and Human Behavior*, transl. R. D. Martin. London: Methuen.
Ward, A. J. W. and Hart, P. J. B. (2005) Foraging benefits of shoaling with familiars. *Animal Behaviour*, **69**, 329–35.
Griesser, M. and Ekman, J. (2005) Nepotistic mobbing behavior in the Siberian Jay, *Perisoreus infaustus*. *Animal Behaviour*, **69**, 345–52.
Grove, N. (1996) *Birds of North America*. Hong Kong: Hugh Lauter Levin Associates, Inc.
Heinrich, B. (1999) *Mind of the Raven: Investigations and Adventures with Wolf-Birds*. New York: HarperCollins.
Poole, J. H. (1996) *Coming of Age with Elephants*. New York: Hyperion Press.
Connor, R. C. and Micklethwaite, D. W. (1994) *The Lives of Whales and Dolphins*. New York: Henry Holt and Co.
MacDonald, D. (1991) *Running with the Fox*. New York: Facts on File.
Barber, T. X. (1993) *The Human Nature of Birds*. New York: St Martin's Press.
Burger, J. (2001) *The Parrot Who Owns Me: The Story of a Relationship*. New York: Villard.
Hauser, M. (2001) *Wild Minds: What Animals Really Think*. London: Penguin.
Cheney, D. L. and Seyfarth, R. M. (1990) *How Monkeys See the World*. Chicago: University of Chicago Press.
Hutto, J. (1995) *Illumination in the Flat Woods: A Season with the Wild Turkey*. New York: Lyons & Burford.
Fraser, B. (1971) *Sitting Duck: A True Story*. London: MacGibbon & Kee.

Chapter 3

Nollman, J. (1986) *Dolphin Dreamtime: Talking to the Animals*. London: Anthony Blond.
Masson, J. M. and McCarthy, S. (1995) *When Elephants Weep: The Emotional Lives of Animals*. New York: Delacorte.
The Gorilla Foundation (2000) [cited August 2005] Available from: http://www.koko.org/world/.
Morris, D. (1990) *Animalwatching: A Field Guide to Animal Behavior*. London: Jonathan Cape.
Herman, L. M. (1986) Cognition and language competencies of bottlenosed dolphins. In: Schusterman, R. J., Thomas, J. A. and

Wood, F. G. (eds.) *Dolphin Cognition and Behavior: a Comparative Approach*. Hillsdale, NJ: Erlbaum, pp. 221–52.

Shettleworth, S. J. (2001) Animal cognition and animal behaviour. *Animal Behaviour*, **61**, 277–86.

Shettleworth, S. J. (1993) Where is the comparison in comparative cognition? *Psychological Science*, **4**, 179–84.

Budiansky, S. (1998) *If a Lion could Talk: Animal Intelligence and the Evolution of Consciousness*. New York: The Free Press.

Tomback, D. F. (1982) Dispersal of whitebark pineseeds by Clark's Nutcracker: a mutualism hypothesis. *Journal of Animal Ecology*, **51**, 451–67.

Downer, J. (2002) *Weird Nature*. London: BBC.

Silverman, A. P. (1978) *Animal Behavior in the Laboratory*. New York: Pica; London: Chapman & Hall.

Masson, J. M. (2004) *The Pig Who Sang to the Moon*. London: Jonathan Cape.

Davis, K. (2001) *More than a Meal: The Turkey in History, Myth, Ritual, and Reality*. New York: Lantern Books.

Davis, K. (1997) *Prisoned Chickens, Poisoned Eggs: An Inside Look at the Modern Poultry Industry*. Summerville, TN: Book Publishing Company.

Rogers, L. J. (1997) *Minds of Their Own: Thinking and Awareness in Animals*. Boulder, CO: Westview Press.

Epstein, R., Lanca, R. P. and Skinner, B. F. (1981) 'Self-awareness' in the pigeon. *Science*, **212**, 695–6.

Marler, P. (1996) Social cognition: are primates smarter than birds? In: Nolan Jr, V. and Ketterson, E. D. (eds.) *Current Ornithology*. New York: Plenum Press, pp. 1–32.

Jarvis, E. D., Güntürkün, O., Bruce, L., Csillag, A., Karten, H., Kuenzel, W., Medina, L., Paxinos, G., Perkel, D. J., Shimizu, T., Striedter, G., Wild, M., Ball, G. F., Dugas-Ford, J., Durand, S., Hough, G., Husband, S., Kubikova, L., Lee, D., Mello, C. V., Powers, A., Siang, C., Smulders, T. V., Wada, K., White, S. A., Yamamoto, K., Yu, J., Reiner, A. and Butler, A. B. (2005) Avian brains and a new understanding of vertebrate brain evolution. *Nature Reviews Neuroscience*, **6**, 151–9.

Weiss, R. (2005) Bird brains get some new names, and new respect. *Washington Post*, 1 February, p. A10.

Davies, C. (2004) How do homing pigeons navigate? They follow roads. *Weekly Telegraph*, 2 May.

Handwerk, B. (2005) Shark facts: Attack stats, record swims, more. *National Geographic News*, 13 June. [cited July 2005] Available from: http://news.nationalgeographic.com/news/2005/06/0613_050613_sharkfacts.html

British Broadcasting Corporation (2003) *Smart Sharks: Swimming with RoboShark*. [television documentary].

Tomkies, M. (1985) *Out of the Wild*. London: Jonathan Cape.

British Broadcasting Corporation (BBC News) (2004) *Crafty Sheep Conquer Cattle Grids*. 30 July. http://news.bbc.co.uk/1/hi/uk/3938591.stm.

Griffin, D. R. (1992) *Animal Minds*. Chicago: University of Chicago Press, p. 260.

Burghardt, G. M. (1991) Cognitive ethology and critical anthropomorphism: a snake with two heads and hognose snakes that play dead. In: Ristau, C. A. (ed.) *Cognitive Ethology: the Minds of Other Animals*. San Francisco: Erlbaum, pp. 53–90.

Downer, J. (2002) *Weird Nature*. London: BBC.

Griffin, D. R. (1984) *Animal Thinking*. Harvard: Harvard University Press.

Kendrick, K. M., Leigh, A. E. and Peirce, J. (2001) Behavioral and neural correlates of mental imagery in sheep using face recognition paradigms. *Animal Welfare*, **10**, 89–101.

Kendrick, K. M., da Costa, A. P., Leigh, A. E., Hinton, M. R. and Peirce, J. W. (2001) Sheep don't forget a face. *Nature*, **414**, 165–6.

Steinhart, P. (1995) *The Company of Wolves*. New York: Alfred A. Knopf.

Van Lawick, H. (1986) *Among Predators and Prey: A Photographer's Reflections on African Wildlife*. San Francisco: Sierra Club Books, p. 71.

Wehner, R., Meier, C. and Zollikofer, C. (2004) The ontogeny of foraging behavior in desert ants, *Cataglyphis bicolor. Ecological Entomology*, **29**, 240–50.

Uhlenbroek, C. (2002) *Talking with Animals*. London: Hodder & Stoughton.

Bekoff, M. (2002) *Minding Animals: Awareness, Emotions, and Heart*. Oxford: Oxford University Press.

Linden, E. (2003) *The Octopus and the Orangutan: New Tales of Animal Intrigue, Intelligence, and Ingenuity*. London: Plume.

British Broadcasting Corporation (2003) *The Life of Mammals*. 2. The Insect Hunters. [television documentary]

Dehnhardt, G., Mauck, B., Hanke, W. and Bleckmann, H. (2001) Hydrodynamic trail-following in harbor seals (*Phoca vitulina*). *Science*, **293**, 102–4.

Eaton, M. D. and Lanyon, S. M. (2003) The ubiquity of avian ultraviolet plumage reflectance. *Proceedings of the Royal Society of London B*, **270**, 1721–6.

Balcombe, J. P. (1990) Vocal recognition of pups by mother Mexican free-tailed bats, *Tadarida brasiliensis mexicana*. *Animal Behaviour*, **39**, 960–6.

Balcombe, J. P. and McCracken, G. F. (1992) Vocal recognition in Mexican free-tailed bats: Do pups recognize mothers? *Animal Behaviour*, **43**, 79–87.

Darwin, C. (1998) *The Expression of the Emotions in Man and Animals*, 3rd edn. (ed. Ekman, P.). London: HarperCollins, p. 50.

Krutch, J. W. (1956) *The Great Chain of Life*. Boston: Houghton Mifflin, p. 106.

Bentham, J. (1789) *An Introduction to the Principles of Morals and Legislation*. London: T. Payne.

Chapter 4

Adams, D. (1984) *So Long, and Thanks for All the Fish*. New York: Crown.

Bekoff, M. and Byers, J. A. (eds.) (1998) *Animal Play: Evolutionary, Comparative, and Ecological Perspectives*. Cambridge: Cambridge University Press, p. xiv.

Groos, M. (1898) *The Play of Animals*. New York: Appleton.

Aldis, O. (1975) *Play Fighting*. New York: Academic Press.

Smith, E. O. (ed.) (1978) *Social Play in Primates*. New York: Academic Press.

Symons, D. (1978) *Play and Aggression: A Study of Rhesus Monkeys*. New York: Columbia University Press.

Fagen, R. M. (1981) *Animal Play Behavior*. New York: Oxford University Press.

Smith, P. K. (1984) *Play in Animals and Humans*. Oxford: Blackwell.

Burghardt, G. (2005) *The Genesis of Animal Play: Testing the Limits*. Cambridge, MA: Bradford Books.

Goodall, J. and Bekoff, M. (2002) *The Ten Trusts: What We Must Do to Care for the Animals We Love*. San Francisco: HarperCollins.

Linden, E. (1999) *The Parrot's Lament: And Other True Tales of Animal Intrigue, Intelligence, and Ingenuity*. New York: Dutton.

Siviy, S. M. (1998) Neurobiological substrates of play behavior: glimpses into the structure and function of mammalian playfulness. In M. Bekoff and J. Byers (eds.) *Animal Play: Evolutionary, Comparative, and Ecological Perspectives*. New York: Cambridge University Press.

Panksepp, J. (2000) The rat will play. In: Bekoff, M. (ed.) *The Smile of a Dolphin: Remarkable Accounts of Animal Emotions*. New York: Discovery Books, pp. 146–7.

Uhlenbroek, C. (2002) *Talking with Animals*. London: Hodder & Stoughton.

Strum, S. (1987) *Almost Human: A Journey into the World of Baboons*. London: Elm Tree Books.

Heinrich, B. (1999) *Mind of the Raven: Investigations and Adventures with Wolf-Birds*. New York: HarperCollins, p. 286.

Pozis-François, O., Zahavi, A. and Zahavi, A. (2004) Social play in Arabian Babblers. *Behaviour*, **141**, 425–50.

Masson, J. M. (2004) *The Pig Who Sang to the Moon*. London: Jonathan Cape.

Morris, D. (1990) *Animalwatching: A Field Guide to Animal Behavior*. London: Jonathan Cape.

McDonnell, S. M. and Poulin, A. (2002) Equid play ethogram. *Applied Animal Behavior Science*, **78**, 263–90.

Bekoff, M. (2004) Wild justice, cooperation, and fair play: minding manners, being nice, and feeling good. In: Sussman, R. and Chapman, A. (eds.) *The Origins and Nature of Sociality*. Chicago: Aldine, pp. 53–79.

Pellis, S. (2002) Keeping in touch: play fighting and social knowledge. In: Bekoff, M., Allen, C. and Burghardt, G. M. (eds.) (2002) *The Cognitive Animal*. Cambridge, MA: MIT Press.

Watson, D. M. and Croft, D. B. (1996) Age-related differences in playfighting strategies of captive male red-necked wallabies (*Macropus rufogriseus banksianus*). *Ethology*, **102**, 336–46.

Hunter, L. (2000) *Cheetah*. Granton-on-Spey: Colin Baxter Photography Ltd.

van den Berg, C. L., Hol, T., Van Ree, J. M., Spruijt, B. M., Everts, H. and Koolhaas, J. M. (1999) Play is indispensable for an adequate development of coping with social challenges in the rat. *Developmental Psychobiology*, **34**, 129–38.

Wolff, R. J. (1981) Solitary and social play in wild *Mus musculus*. *Journal of Zoology, London*, **195**, 405–12.

Walker, C. and Byers, J. A. (1991) Heritability of locomotor play in house mice, *Mus domesticus*. *Animal Behaviour*, **42**, 891–7.
Marashi, V., Barnekow, A., Ossendorf, E. and Sachser, N. (2003) Effects of different forms of environmental enrichment on behavioral, endocrinological, and immunological parameters in male mice. *Hormones & Behavior*, **43**, 281–92.
Evershed, S. (1930) Ravens flying upside down. *Nature*, **126**, 956–7.
Tåning, A. V. (1931) Ravens flying upside-down. *Nature*, **127**, 856.
Moffett, A. T. (1984) Ravens sliding in snow. *British Birds*, **77**, 321–2.
Heinrich, B. (1999) *Mind of the Raven: Investigations and Adventures with Wolf-Birds*. New York: HarperCollins.
Heinrich, B. and Smolker, R. (1998) Play in common ravens (*Corvus corax*). In: Bekoff, M. and Byers, J. A. (eds.) *Animal Play: Evolutionary, Comparative, and Ecological perspectives*. Cambridge: Cambridge University Press, pp. 27–44.
Key, Sheila (2003) Personal communication, September.
British Broadcasting Corporation (2003) *Wild Down Under* [television documentary].
Burghardt, G. M. (2005) *The Genesis of Animal Play: Testing the Limits*. Cambridge, MA: Bradford Books.
Karasawa, K. (1992) Playing crows. *Urban Birds*, **9**, 31 (in Japanese).
Karasawa, K. (1996) *Karasu wa Tensai!* (Crows are talented.) Tokyo: Goma Shobo (in Japanese).
Burger, J. (2001) *The Parrot Who Owns Me: The Story of a Relationship*. New York: Villard.
Marsden, E. (2003) Personal communication, September.
Ortega, J. C. and Bekoff, M. (1987) Avian play: comparative evolutionary and developmental trends. *Auk*, **104**, 338–41.
Chapman, E. (2003) Personal communication, September.
Caffrey, C. (2001) Goal directed use of objects by American Crows. *Wilson Bulletin*, **113**, 114–15.
Kramer, P. (2003) Personal communication, September.
Hope, S. (2003) Personal communication, September.
Grunwald, J.-E., Schörnich, S. and Wiegrebe, L. (2004) Classification of natural textures in echolocation. *Proceedings of the National Academy of Sciences*, **101**, 5670–4.
Simmons, R. E. and Mendelsohn, J. M. (1993) A critical review of cartwheeling flights of raptors. *Ostrich*, **64**, 13–24.
Goodall, J. (1986) *The Chimpanzees of Gombe: Patterns of Behavior*. Cambridge, MA: Belknap Press.

Moss, C. and Colbeck, M. (1992) *Echo of the Elephants: the Story of an Elephant Family*. London: BBC Books.

Pozis-Francois, O., Zahavi, A. and Zahavi, A. (2004) Social play in Arabian Babblers. *Behavior*, **141**, 425–50.

Estes, R. D. (1991) *The Behavior Guide to African Mammals*. Berkeley, CA: University of California Press.

Watson, D. M. (1998) Kangaroos at play: play behavior in the Macropodoidea. In: Bekoff, M. and Byers, J. A. (eds.) *Animal Play: Evolutionary, Comparative, and Ecological Perspectives*. Cambridge: Cambridge University Press, pp. 61–95.

Tomkies, M. (1985) *Out of the Wild*. London: Jonathan Cape.

Thurston, M. (2003) Personal communication, June.

McCowan, B., Marino, L., Vance, E., Walke, L. and Reiss, D. (2000) Bubble ring play of bottlenose dolphins (*Tursiops truncatus*): implications for cognition. *Journal of Comparative Psychology*, **114**, 98–106.

Eberst, A. (2004) Ring of bright air. *New Scientist*, **181**, 81.

Connor, R. C. and Micklethwaite, D. W. (1994) *The Lives of Whales and Dolphins*. New York: Henry Holt and Co.

Herman, L. M. (1986) Cognition and language competencies of bottlenosed dolphins. In: Schusterman, R.J., Thomas, J.A. and Wood, F. G. (eds.) *Dolphin Cognition and Behavior: a Comparative Approach*. Hillsdale, NJ: Erlbaum, pp. 221–2.

Roberts, B. (1934) Notes on the birds of central and south-east Iceland with special reference to food habits. *Ibis*, **13**, 239–64.

Thompson, A. L. (1964) *A New Dictionary of Birds*. New York: McGraw-Hill.

Stoner, E. A. (1947) Anna hummingbird at play. *Condor*, **49**, 36.

Shane, S. H., Wells, R. S. and Wursig, B. (1986) Ecology, behavior and social organization of the bottlenose dolphin: a review. *Marine Mammal Science* **2**(1), 34–63.

Howard, C. J. (1995) *Dolphin Chronicles*. New York: Bantam.

Heyman, K. (2003) Personal communication, May.

Tsipoura, N. (2005) Personal communication, June.

Burghardt, G. M. (2004) Play: how evolution can explain the most mysterious behavior of all. In: Moya, A. and Font, E. (eds.) *Evolution: From Molecules to Ecosystems*. Oxford: Oxford University Press, pp. 231–46.

Kramer, M. and Burghardt, G. M. (1998) Precocious courtship and play in emydid turtles. *Ethology*, **104**, 338–56.

Burghardt, G. M. (1998) The evolutionary origins of play revisited: lessons from turtles. In Bekoff, M. and Byers, J. A. (eds.) (1998) *Animal Play: Evolutionary, Comparative, and Ecological Perspectives*. Cambridge: Cambridge University Press, pp. 1–26.

Marcum, H. and Dawson, C. (2004) Back scratching and enrichment in sea turtles. In Bekoff, M. (ed.) *Encyclopedia of Animal Behavior*. Westport, CT: Greenwood Press, pp. 1093–5.

Lazell, J. D. and Spitzer, N. C. (1977) Apparent play behavior in an American alligator. *Copeia*, **1977**(1), 188–9.

Kuba, M., Meisel, D. V., Byrne, R. A., Griebel, U. and Mather, J. A. (2003) Looking at play in *Octopus vulgaris*. In: Warnke, K., Keupp, H. and Boletzky, S. V. (eds.) Coleoid cephalopods through time. *Berliner Paläontologische Abhandlungen*, **3**, 163–9.

Linden, E. (2003) *The Octopus and the Orangutan: New Tales of Animal Intrigue, Intelligence, and Ingenuity*. London: Plume.

Brown, S. L. (1994) Animals at play. *National Geographic*, December, p. 30.

Young, R. (2003) *The Secret Life of Cows*. Preston: Farming Books and Videos Ltd.

Van Lawick, H. (1986) *Among Predators and Prey: A Photographer's Reflections on African Wildlife*. San Francisco: Sierra Club Books.

Crumley, J. (1992) *Waters of the Wild Swan*. London: Jonathan Cape.

Wursig, B. (2000) In a party mood. In: Bekoff, M. (ed.) (2000) *The Smile of a Dolphin: Remarkable Accounts of Animal Emotions*. New York: Discovery Books, pp. 188–91.

Mather, J. (1992) Underestimating the octopus. In: Davis, H. and Balfour, D. (eds.) *The Inevitable Bond: Examining Scientist–Animal Interactions*. New York: Cambridge University Press, pp. 240–9.

de Waal, F. (2005) Suspicious minds. *New Scientist*, **186**, 48.

Bell, N. J. and Bell, R. W. (1993) *Adolescent Risk Taking*. Newbury Park, CA: Sage.

Northcutt, W. (2002) *The Darwin Awards II: Unnatural Selection*. New York: Dutton.

de Waal, F. (1996) *Good Natured: The Origins of Right and Wrong in Humans and Other Animals*. Cambridge, MA: Harvard University Press.

Steinhart, P. (1995) *The Company of Wolves*. New York: Alfred A. Knopf.

Adams, R. G. (1972) *Watership Down*. London: Penguin, p. 69.

Chapter 5

Krutch, J. W. (1969) *The Best Nature Writing of Joseph Wood Krutch.* New York: William Morrow & Co.

Hollingham, R. (2004) In the realm of your senses. *New Scientist*, **181**, 40–3.

Johnston, D. S. and Fenton, M. B. (2001) The diets of pallid bats (*Antrozous pallidus*): variability at individual and population levels. *Journal of Mammalogy*, **82**, 362–73.

Laska, M., Hernandez Salazar, L. T. and Rodriguez Luna, E. (2000) Food preferences and nutrient composition in captive spider monkeys, *Ateles geoffroyi. International Journal of Primatology*, **21**, 671–83.

Pollan, M. (2001) *The Botany of Desire: A Plant's-eye View of the World.* New York: Random House.

Cabanac, M. (2005) The experience of pleasure in animals. in McMillan, F. D. (ed.) *Mental Health and Well-being in Animals.* Ames: Iowa State University Press, pp. 29–46.

Panksepp, J. (1998) *Affective Neuroscience.* Oxford: Oxford University Press.

Cabanac, M. and Johnson, K. G. (1983) Analysis of a conflict between palatability and cold exposure in rats. *Physiology & Behavior*, **31**, 249–53.

Cabanac M. (1985) Strategies adopted by juvenlie lizards foraging in a cold environment. *Physiological Zoology*, **58**, 262–71.

Balaskó, M. and Cabanac, M. (1998) Behavior of juvenile lizards (*Iguana iguana*) in a conflict between temperature regulation and palatable food. *Brain, Behavior and Evolution*, **52**, 257–62.

Young, R. (2003) *The Secret Life of Cows.* Preston: Farming Books and Videos Ltd.

Galef Jr, B. G. and Whiskin, E. E. (2003) Preference for novel flavors in adult Norway rats (*Rattus norvegicus*). *Journal of Comparative Psychology*, **117**, 96–100.

Patterson, F. (2002) Koko Foundation. Personal communication, September.

Scutro, A. (2004) Chasing the killers: killer whale researchers document the recent violence of Monterey Bay. *Monterey County Weekly.* [cited August 2005] Available from: http://www.montereycountyweekly.com/issues/Issue.06-03-2004/cover/Article.coverstory.

Heinrich, B. (1999) *Mind of the Raven: Investigations and Adventures with Wolf-Birds.* New York: HarperCollins.

Haslam, C. (2004) Ooh, the jungle VIPs. *The Sunday Times*, 1 August (section 4, page 6).

Masson, J. M. and McCarthy, S. (1995) *When Elephants Weep: The Emotional Lives of Animals*. New York: Delacorte.

Linden, E. (2003) *The Octopus and the Orangutan: New Tales of Animal Intrigue, Intelligence, and Ingenuity*. London: Plume.

Burger, J. (2001) *The Parrot Who Owns Me: The Story of a Relationship*. New York: Villard.

Berridge, K. (1996) Food reward: brain substrates of wanting and liking. *Neuroscience and Biobehavioral Reviews*, **20**, 1–25.

Neuringer, A. (1969) Animals respond for food in the presence of free food. *Science*, **166**, 399–401.

Spruijt, B. M., van den Bos, R. and Pijlman, F. T. A. (2001) A concept of welfare based on reward evaluating mechanisms in the brain: anticipatory behavior as an indicator for the state of reward systems. *Applied Animal Behavior Science*, **72**, 145–71.

Dawkins, M. S. (1998) Evolution and animal welfare. *The Quarterly Review of Biology*, **73**, 305–28.

Olsson, A. S. and Dahlborn, K. (2002) Improving housing conditions for laboratory mice: a review of 'environmental enrichment.' *Laboratory Animals*, **36**, 243–70.

Tinklepaugh, O. L. (1928) An experimental study of representative factors in monkeys. *Journal of Comparative Psychology*, **8**, 197–236.

de Waal, F. (1996) *Good Natured: The Origins of Right and Wrong in Humans and Other Animals*. Cambridge, MA: Harvard University Press.

Da Costa, A. P. C., Leigh, A. E., Man, M.-S. and Kendrick, K. M. (2004) Face pictures reduce behavioral, autonomic, endocrine and neural indices of stress and fear in sheep. *Proceedings of the Royal Society B: Biological Sciences*, **271**, 2077–84.

Strieck, F. (1924) Sensory, neural and behavioral physiology. *Z. vergleich Physiol.*, **2**, 122–54.

Krinner, M. (1935) Über die Geschmacksempfindlichkeit der Elritzen. *Z. vergleich Physiol.*, **21**, 319–42.

Brown, M. E. (ed.) (1957) *The Physiology of Fishes. Vol II: Behavior.* New York: Academic Press.

Herrick, C. J. (1903) The organ and sense of taste in fishes. *Bullettin of the US Fish Commission*, **22**, 237–71.

Kasumyan, A. O. and Doeving, K. B. (2003) Taste preferences in fishes. *Fish and Fisheries*, **4**, 289–347.

Chapter 6

Bagemihl, B. (1999) *Biological Exuberance: Animal Homosexuality and Natural Diversity*. London: Profile Books Ltd.

Judson, O. (2003) *Dr Tatiana's Sex Advice to All Creation: The Definitive Guide to the Evolutionary Biology of Sex*. London: Vintage, p. 59.

Burger, J. (2001) *The Parrot Who Owns Me: The Story of a Relationship*. New York: Villard.

British Broadcasting Corporation (2003) *The Truth about Gay Animals*. [television documentary].

Hrdy, S. B. (1988) The primate origins of human sexuality. In: *The Evolution of Human Sexuality*. Oxford: Oxford University Press.

Lloyd, E. A. (2005) *The Case of the Female Orgasm: Bias in the Science of Evolution*. Cambridge, MA: Harvard University Press.

Chevalier-Skolnikoff, S. (1974) The ontogeny of communication in the stumptail macaque (*Macaca arctoides*). *Contributions to Primatology*, **2**, 1–166.

Dixson, A. F. (1998) *Primate Sexuality: Comparative Studies of the Prosimians, Monkeys, Apes, and Human Beings*. Oxford: Oxford University Press.

Masson, J. M. and McCarthy, S. (1995) *When Elephants Weep: The Emotional Lives of Animals*. New York: Delacorte.

Gould, S. J. and Lewontin, R. (1979) The spandrels of San Marco and the Panglossian paradigm: a critique of the adaptionist program. *Proceedings of the Royal Society of London B*, **205**, 581–98.

Vachon, P., Simmerman, N., Zahran, A. R. and Carrier, S. (2000) Increases in clitoral and vaginal blood flow following clitoral and pelvic plexus nerve stimulations in the female rat. *International Journal of Impotence Research*, **12**(1), 53–7.

Munarriz, R., Kim, S. W., Kim, N. N., Traish, A. and Goldstein, I. (2003) A review of the physiology and pharmacology of peripheral (vaginal and clitoral) female genital arousal in the animal model. *Journal of Urology*, **170**(2 Pt 2), S40–4; discussion S44–5.

de Waal, F. (1982) *Chimpanzee Politics: Power and Sex Among Apes*. New York: Harper & Row.

Fouts, R. and Mills, S. T. (1997) *Next of Kin: What Chimpanzees Have Taught Me About Who We Are*. New York: William Morrow.

Howard, C. J. (1995) *Dolphin Chronicles*. New York: Bantam.

Wursig, B. (2000) In a party mood. In: Bekoff, M. (ed.) (2000) *The Smile of a Dolphin: Remarkable Accounts of Animal Emotions*. New York: Discovery Books, pp. 188–91.

Knudtson, P. (1996) *Orca: Visions of the Killer Whale*. San Francisco: Sierra Club Books.

Van der Harst, J. E., Fermont, P. C. J., Bilstra, A. E. and Spruijt, B. M. (2003) Access to enriched housing is rewarding to rats as reflected by their anticipatory behavior. *Animal Behaviour*, **66**, 493–504.

Poduschka, W. (1981) Abnormes Sexualverhalten Zusammengehaltener, Weiblicher *Hemiechinus auritus syriacus* (Insectivora: Erinaceinae) [Abnormal sexual behavior of confined female *Hemiechinus auritus syriacus*]. *Bejdragen tot de Dierkunde*, **51**, 81–8.

Pepperberg, I. (2000) *The Alex Studies: Cognitive and Communicative Abilities of Grey Parrots*. Harvard: Harvard University Press.

Gaston, T. and Kampp, K. (1994) Thick-billed murre masturbating on grass clump. *Pacific Seabirds*, **21**, 30.

Winterbottom, M., Burke, T. and Birkhead, T. R. (2001) The phalloid organ, orgasm and sperm competition in a polygynandrous bird: the red-billed buffalo weaver (*Bubalornis niger*). *Behavioral Ecology and Sociobiology*, **50**, 474–82.

Wikelski, M. and Baurle, S. (1996) Pre-copulatory ejaculation solves time constraints during copulations in marine iguanas. *Proceedings of the Royal Society of London B*, **263**, 439–44.

Chapter 7

Butler, S. (1912) *Notebooks*. New York: EP Dutton & Company (1951), p. 154.

Darwin, C. (1998) *The Expression of the Emotions in Man and Animals*, 3rd edn (ed. Ekman, P.). London: HarperCollins.

Lawrence, D. H. (1923) *Birds, Beasts and Flowers: Poems*. London: Martin Secker.

Burghardt, G. M. (1998) The evolutionary origins of play revisited: lessons from turtles. In Bekoff, M. and Byers, J. A. (eds.) (1998) *Animal Play: Evolutionary, Comparative, and Ecological Perspectives*. Cambridge: Cambridge University Press, pp. 1–26.

Uhlenbroek, C. (2002) *Talking with Animals*. London: Hodder & Stoughton, p. 149.

Burghardt, G. M. (2004) Iguana research: looking back and looking ahead. In Alberts, A. D., Carter, R. L., Hayes, W. K. and Martins, E. P. (eds.) *Iguanas: Biology and Conservation*. Berkeley, CA: University of California Press, pp. 1–12.

Connor, R. C. and Micklethwaite, D. W. (1994) *The Lives of Whales and Dolphins*. New York: Henry Holt and Co.

Linden, E. (2003) *The Octopus and the Orangutan: New Tales of Animal Intrigue, Intelligence, and Ingenuity*. London: Plume.

Nollman, J. (1986) *Dolphin Dreamtime: Talking to the Animals*. London: Anthony Blond.

Knudtson, P. (1996) *Orca: Visions of the Killer Whale*. San Francisco: Sierra Club Books.

Kilcommons, B. and Wilson, S. (2003) A cat can steal your heart. *Parade Magazine*, 1 June.

Young, R. (2003) *The Secret Life of Cows*. Preston: Farming Books and Videos Ltd.

Keverne, E. B. (1992) Primate social relationships their determinants and consequences. *Advances in the Study of Behavior*, **21**, 1–36.

Feh, C. and de Mazières, J. (1993) Grooming at a preferred site reduces heart rate in horses. *Animal Behaviour*, **46**, 1191–4.

Strum, S. (1987) *Almost Human: A Journey into the World of Baboons*. London: Elm Tree Books.

Alcock, J. (1989) *Animal Behavior*, 4th edn. Sunderland, MA: Sinauer.

Burger, J. (2001) *The Parrot Who Owns Me: The Story of a Relationship*. New York: Villard.

Tobias, J. (2005) A trumpeter's tale. *BBC Wildlife*, July, 36–9.

Goodwin, D. (1983) *Pigeons and Doves of the World*, 3rd edn. London: British Museum (Natural History).

Heinrich, B. (1999) *Mind of the Raven: Investigations and Adventures with Wolf-Birds*. New York: HarperCollins, p. 286.

Harrison, C. J. O. (1965) Allopreening as agonistic behavior. *Behavior*, **24**, 161–209.

Barber, T. X. (1993) *The Human Nature of Birds*. New York: St Martin's Press.

Tomkies, M. (1985) *Out of the Wild*. London: Jonathan Cape, p. 111.

Attenborough, D. (1990) *The Trials of Life*. Boston: Little, Brown and Company.

Bowers, B. and Burghardt, G. (1992) The scientist and the snake: relationships with reptiles. In: Davis, H. and Balfour, D. (eds.) *The Inevitable Bond: Examining Scientist–Animal Interactions*. New York: Cambridge University Press, p. 250–63.

Deeble, M. and Stone, V. (2001) Kenya's Mzima Spring, *National Geographic*, **200**, 32–47.

Deeble, M. (2003) Hippo heaven. *BBC Wildlife*, **21**, 42–9.

Downer, J. (2002) *Weird Nature*. London: BBC.

Van Lawick, H. (1986) *Among Predators and Prey: A Photographer's Reflections on African Wildlife*. San Francisco: Sierra Club Books.

Morgan, J. (2004) Personal communication, February.

Times of India (2003) Leopard–cow bond thrills villagers. 14 May. http://www.hvk.org/articles/0503/208.html

Barboza, D. (2003) Animal welfare's unexpected allies. *The New York Times*, 25 June.

Bagemihl, B. (1999) *Biological Exuberance: Animal Homosexuality and Natural Diversity*. London: Profile Books Ltd.

Burgdorf, J. and Panksepp, J. (2001) Tickling induces reward in adolescent rats. *Physiology & Behavior*, **72**, 167–73.

Randerson, J. (2003) Marine census reveals depth of ignorance. *New Scientist*, **180**, 14.

Laland, K., Brown, C. and Krause, J. (2003) Learning in fishes: an introduction. *Fish and Fisheries*, **4**, 199–202.

Brown, C. (2004) Clever fish: not just a pretty face. *New Scientist*, **182**, 42–3.

Potts, G. W. (1973) The ethology of *Labroides dimidiatus* on Aldabra. *Animal Behaviour*, **21**, 250–91.

Bshary, R. and Wuerth, M. (2001) Cleaner fish *Labroides dimidiatus* manipulate client reef fish by providing tactile stimulation. *Proceedings of the Royal Society of Londons B: Biological Sciences*, **268**, 1495–501.

Tebbich, S., Bshary, R. and Grutter, A. S. Cleaner fish *Labroides dimidiatus* recognize familiar clients. *Animal Cognition*, **5**, 139–45.

Bshary, R. and Shaeffer, D. (2002) Choosy reef fish select cleaner fish that provide high-quality service. *Animal Behaviour*, **63**, 557–64.

Morris, D. (1984) *The Naked Ape*. New York: Dell.

Becklund, J. (1999) *Summers with the Bears: Six Seasons in the North Woods*. New York: Hyperion.

Morris, D. (1990) *Animalwatching: A Field Guide to Animal Behavior*. London: Jonathan Cape.

Smith, J. (1981) Senses and communication. In: Berry, R. (ed.) *Biology of the House Mouse*. London: Academic Press.

Paling, J. (1979) *Squirrel on My Shoulder*. London: BBC.

Website of photographers Nicole and André Brunsperger. [cited Feb 2004] Available from: http://www.stock-image.org/resultgb.php3?langue=en&keys=lion&actions=suite&index_debut=8&nombre_photo=47

Clark, R. (2004) Personal communication, April.

McGowan, K. J. and Woolfenden, G. E. (1986) Aerial rain bathing by Common Nighthawks. *The Wilson Bulletin*, **98**, 612–13.

Chapter 8

Hamley, Colonel E. B. (1872) *Our Poor Relations*. Edinburgh: William Blackwood & Sons.

Black, J. M. (ed.) (1996) *Partnerships in Birds: The Study of Monogamy*. Oxford: Oxford University Press.

Bekoff, M. (ed.) (2000) *The Smile of a Dolphin: Remarkable Accounts of Animal Emotions*. New York: Discovery Books.

Goodall, J. (1986) *The Chimpanzees of Gombe: Patterns of Behavior*. Cambridge, MA: Belknap Press.

Goodall, J. and Bekoff, M. (2002) *The Ten Trusts: What We Must do to Care for the Animals We Love*. San Francisco: HarperCollins.

Heinrich, B. (1999) *Mind of the Raven: Investigations and Adventures with Wolf-Birds*. New York: HarperCollins.

Masson, J. M. (2004) *The Pig Who Sang to the Moon*. London: Jonathan Cape.

Hartshorne, C. (1973) *Born to Song: An Interpretation and World Survey of Bird Song*. Bloomington: Indiana University Press.

Tomkies, M. (1985) *Out of the Wild*. London: Jonathan Cape, p. 176.

Carter, C. S. (1998) Neuroendocrine perspectives on social attachment and love. *Psychoneuroendocrinology*, **23**, 779–818.

Panksepp, J. (1998) *Affective Neuroscience*. Oxford: Oxford University Press.

Rose, N. A. (2000) Giving a little latitude. In: Bekoff, M. (ed.) (2000) *The Smile of a Dolphin: Remarkable Accounts of Animal Emotions*. New York: Discovery Books, pp. 32–3.

Wrangham, R. A. (2000) Making a baby. In: Bekoff, M. (ed.) (2000) *The Smile of a Dolphin: Remarkable Accounts of Animal Emotions*. New York: Discovery Books, pp. 34–5.

Masson, J. M. and McCarthy, S. (1995) *When Elephants Weep: The Emotional Lives of Animals*. New York: Delacorte, p. 73.

Rowell, T. (1972) *The Social Behavior of Monkeys*. London: Penguin Books.

Fricke, H. W. (1979) Mating system, resource defense and sex change in the anemonefish *Amphiprion akallopisos*. *Zeitschrift für Tierpsychologie*, **50**, 313–26.

Taborsky, M. (1984) Broodcare helpers in the cichlid fish *Lamprologus brichardi*: their costs and benefits. *Animal Behaviour*, **32**, 1236–52.

Koenig, W. D. and Dickinson, J. L. (eds.) (2004) *Ecology and Evolution of Cooperative Breeding in Birds*. Cambridge: Cambridge University Press.

French, J. A. (1996) *Cooperative Breeding in Mammals.* Cambridge: Cambridge University Press.

Grove, N. (1996) *Birds of North America.* Hong Kong: Hugh Lauter Levin Associates, Inc.

Kowalski, G. (1999) *The Souls of Animals*. Walpole, NH: Stillpoint.

Hannan, J. *et al.* (2005) Stunned reactions (letters to the editor). *Birds*, **20**, 4.

Temerlin, M. K. (1975) *Lucy: Growing Up Human: A Chimpanzee Daughter in a Psychotherapist's Family.* Palo Alto: Science & Behavior Books.

Burger, J. (2001) *The Parrot Who Owns Me: The Story of a Relationship.* New York: Villard.

CBS Broadcasting Incorporated (2005) Polly caught a killer. [cited Dec 2005] Available from http://www.cbsnews.com/stories/2003/02/19/national/main541186.shtml

Linden, E. (1999) *The Parrot's Lament: And Other True Tales of Animal Intrigue, Intelligence, and Ingenuity.* New York: Dutton.

Chapter 9

Kowalski, G. (1999) *The Souls of Animals*. Walpole, NH: Stillpoint, p. 139.

Trivedi, B. P. (2002) Zoos use new tricks to stimulate animals. *National Geographic News*, 27 August.

Downer, J. (2002) *Weird Nature.* London: BBC.

Pollan, M. (2001) *The Botany of Desire: A Plant's-eye View of the World.* New York: Random House.

Linden, E. (1999) *The Parrot's Lament: And Other True Tales of Animal Intrigue, Intelligence, and Ingenuity.* New York: Dutton.

Hill, J. O., Pavlik, E. J., Smith, G. L. III, Burghardt, G. M. and Coulson, P. B. (1976) Species-characteristic responses to catnip by undomesticated felids. *Journal of Chemical Ecology*, **2**, 239–53.

Young, R. (2003) *The Secret Life of Cows.* Preston: Farming Books and Videos Ltd.

Meredith, M. (2004) *Elephant Destiny: Biography of an Endangered Species in Africa.* New York: PublicAffairs.

Heinrich, B. (1999) *Mind of the Raven: Investigations and Adventures with Wolf-Birds*. New York: HarperCollins, p. 286.

Morris, D. (1990) *Animalwatching: A Field Guide to Animal Behavior.* London: Jonathan Cape.

Abate, F. R. (2002) *The Oxford Pocket American Dictionary of Current English.* New York: Oxford University Press.

Goodall, J. (1986) *The Chimpanzees of Gombe: Patterns of Behavior.* Cambridge, MA: Belknap Press.

de Waal, F. (1982) *Chimpanzee Politics: Power and Sex Among Apes.* New York: Harper & Row, p. 26.

Köhler, W. (1925) *The Mentality of Apes.* New York: Harcourt Brace and World [quoted in Goodall 1986, p. 242].

Crane, S. (2002) Quoted in: Scully, M. *Dominion: The Power of Man, the Suffering of the Animals, and the Call to Mercy.* New York: St Martin's Press.

Griffin, D. R. (1992) *Animal Minds.* Chicago: University of Chicago Press.

Humphrys, J. (2003) Happy cows make a big difference to our health. *The Sunday Times,* 9 November, p. 1-21.

Poole, J. H. (1996) *Coming of Age with Elephants.* New York: Hyperion Press.

Burger, J. (2001) *The Parrot Who Owns Me: The Story of a Relationship.* New York: Villard.

Fox, H. M. (1952) *The Personality of Animals.* London: Penguin, p. 67.

Marin, M. (1997) On the behavior of the black swift. *Condor,* **99**, 514–19.

Collins, R. J. and Jefferson, D. R. (1992) The evolution of sexual selection and female choice. In Varela, F. J. and Bourgine, P. (eds.) *Toward a Practice of Autonomous Systems: Proceedings of the First European Conference on Artificial Life.* Cambridge, MA: MIT Press/Bradford Books, pp. 327–36.

Barber, T. X. (1993) *The Human Nature of Birds.* New York: St Martin's Press.

Hunt, S., Bennett, A. T. D., Cuthill, I. C. and Griffiths, R. (1998) Blue tits are ultraviolet tits. *Proceedings of the Royal Society of London B, Biological Sciences,* **265**, 451–5.

Mather, J. A. (2004) Behavioral physiology: color vision in animals. In: Bekoff, M. (ed.) *Encyclopedia of Animal Behavior.* Westport, CT: Greenwood Press, p. 92.

Dalton, R. (2004) True colors. *Nature,* **428**, 596–7.

Grove, N. (1996) *Birds of North America.* Hong Kong: Hugh Lauter Levin Associates, Inc.

Borgia, G. (2004) Bowerbirds. In Bekoff, M. (ed.) *Encyclopedia of Animal Behavior.* Westport, CT: Greenwood Press, pp. 881–3.

von Frisch, K. (1974) *Animal Architecture.* New York: Harcourt Brace Jovanovich.

Turner, F. (1991) *Beauty: The Value of Values.* Charlottesville, VA: University Press of Virginia, p. 40.

Skutch, A. (2000) Singing the praises of family. In: Bekoff, M. (ed.) *The Smile of a Dolphin: Remarkable Accounts of Animal Emotions*. New York: Discovery Books, pp. 52–3.

Howard, L. (1956) *Birds as Individuals*. London: Collins.

Rothenberg, D. (2005) *Why Birds Sing: A Journey Through the Mystery of Bird Song*. New York: Basic Books.

Hartshorne, C. (1973) *Born to Song: An Interpretation and World Survey of Bird Song*. Bloomington: Indiana University Press. pp. 53–4, 153–4.

West M. J., King A. P., Goldstein M. H. in press. Singing, socializing, and the music effect. In: Marler, P., Slabbekoorn, H. and Hope, S. (eds.) *Nature's Music: The Science of Bird Song*. Ithaca: Cornell University Press, pp. 374–87.

Mapes, J. (2001) Do animals have an innate sense of music? 5 January [cited March 2004] Available from: http://news.nationalgeographic.com/news/2001/01/0105biomusic.html.

Hauser, M. (2001) *Wild Minds: What Animals Really Think*. London: Penguin, p. 195.

Burger, J. (2001) *The Parrot Who Owns Me: The Story of a Relationship*. New York: Villard.

Watanabe, S. and Sato, K. (1999) Discriminative stimulus properties of music in Java sparrows. *Behavioral Processes*, **47**, 53–7.

Porter, D. and Neuringer, A. (1984) Musical discrimination by pigeons. *Journal of Experimental Psychology: Animal Behavior Processes*, **10**, 138–48.

Poli, M and Previde, E. P. (1991) Discrimination of musical stimuli by rats (*Rattus norvegicus*). *International Journal of Comparative Psychology*, **5**, 7–18.

Hulse, S. H., Humpal, J. and Lynx, J. (1984) Discrimination and generalization of rhythmic and arhythmic sound patterns by European Starlings (*Sturnus vulgaris*). *Music Perception*, **1**, 442–64.

Hulse, S. H., Bernard, D. J. and Braaten, R. F. (1995) Auditory discrimination of chord-based spectral structures by European Starlings (*Sturnus vulgaris*). *Journal of Experimental Psychology: General*, **124**, 409–23.

Rauscher, F. H., Robinson, K. D. and Jens, J. J. (1998) Improved maze learning through early music exposure in rats. *Neurological Research*, **20**, 427–31.

Phillips, H. (2003) Five key questions about pleasure. *New Scientist*, 11 October, 41–3.

Watanabe, S., Sakamoto, J. and Wakita M. (1995) Pigeons' discrimination of paintings by Monet and Picasso. *Journal of the Experimental Analysis of Behavior,* **63**, 165–74.

Provine, R. (2000) *Laughter: A Scientific Investigation*. London: Faber & Faber.

Cardoso, S. H. (2000) Our ancient laughing brain. *Cerebrum*, **2**.

The Gorilla Foundation (2002) *Penny's Journal*, 26 September [cited May 2003] Available from: http://koko.org/world/journal.phtml?offset=11.

Panksepp, J. (2005) Beyond a joke: From animal laughter to human joy? *Science*, **308**, 62–3.

Blumberg, M. S., & Sokoloff, G. (2001) Do infant rats cry? *Psychological Review*, **108**, 83–95.

de Waal, F. (1996) *Good Natured: The Origins of Right and Wrong in Humans and Other Animals*. Cambridge, MA: Harvard University Press.

Bodon, M. (2002) Fundamentals of fun. *Newsday*, 12 February.

Panksepp, J. and Burgdorf, J. (2003) 'Laughing' rats and the evolutionary antecedents of human joy? *Physiology & Behavior,* **79**, 533–47.

Panksepp, J. (2000) The riddle of laughter: neural and psycho-evolutionary underpinnings of joy. *Current Directions in Psychological Science*, **9**, 183–6.

Spa Magazine (2003) Seattle bimonthly, April, p. 20.

Masson, J. M. (2004) *The Pig Who Sang to the Moon*. London: Jonathan Cape.

Grimes, W. (2002) *My Fine Feathered Friend*. New York: North Point Press.

Bekoff, M. (ed.) (2000) *The Smile of a Dolphin: Remarkable Accounts of Animal Emotions*. New York: Discovery Books.

Chapter 10

Bryson, B. (2003) *A Short History of Nearly Everything*. London: Doubleday.

Mayr, E. (2002) *What Evolution Is*. London: Phoenix.

Burghardt, G. M. (2005) *The Genesis of Animal Play: Testing the Limits*. Cambridge, MA: Bradford Books.

Brown, C. (2004) Clever fish: Not just a pretty face. *New Scientist*, **182**, 42–3.

Cabanac, M. (2005) The experience of pleasure in animals. In McMillan, F. D. (ed.) *Mental Health and Well-being in Animals*. Ames: Iowa State University Press, pp. 29–46.

Renbourn, E. T. (1960) Body temperature and the emotions. *Lancet*, **2**, 475–6.

Gotsev, T. and Ivanov, A. (1962) Psychogenic elevation of body temperature. *Proceedings of the International Union of Physiological Science*, **2**, 501.

Briese, E. and deQuijada, M. G. (1970) Colonic temperature of rats during handling. *Acta Physiologica, Pharmacologica et Therapeutica Latinoamericana*, **20**, 97–102.

Cabanac, A. and Cabanac, M. (2000) Heart rate response to gentle handling of frog and lizard. *Behavioral Processes*, **52**, 89–95.

Chandroo, K. P., Yue, S. and Moccia, R. D. (2004) An evaluation of current perspectives on consciousness and pain in fishes. *Fish and Fisheries*, **5**, 281–95.

Sneddon, L. U., Braithwaite, V. A. and Gentle, M. J. (2003) Do fishes have nociceptors? Evidence for the evolution of a vertebrate sensory system. *Proceedings of the Royal Society of London B*, **270**, 1115–21.

Ehrensing, R. H., Michell, G. F. and Kastin, A. J. (1982) Similar antagonism of morphine analgesia by MIF-1 and naxolone in *Carassius auratus*. *Pharmacology, Biochemistry and Behavior*, **17**, 757–61.

Beukema, J. J. (1970a) Angling experiments with carp (*Cyprinus carpio* L.). II. Decreased catchability through one trial learning. *Netherlands Journal of Zoology*, **19**, 81–92.

Beukema, J. J. (1970b) Acquired hook avoidance in the pike *Esox lucius* L. fished with artificial and natural baits. *Journal of Fish Biology*, **2**, 155–60.

Laland, K., Brown, C. and Krause, J. (2003) Learning in fishes: an introduction. *Fish and Fisheries*, **4**, 199–202.

Matthews, R. (2004) Fast-learning fish have memories that put their owners to shame. *The Sunday Telegraph*, 3 October, p. 12.

Aronson, L. R. (1972) Further studies on orientation and jumping behavior in the gobiid fish, *Bathygobius soporator*. *Annals of the New York Academy of Science*, **188**, 378–92.

Dugatkin, L. A. (2000a) I'll have what she's having. In: Bekoff, M. (ed.) *The Smile of a Dolphin: Remarkable Accounts of Animal Emotions*. New York: Discovery Books, pp. 42–3.

Dugatkin, L. A. (2000b) I'll have what she just had: love in the guppy: (*Poecilia reticulata*). In: Bekoff, M. (ed.) *A Passionate Nature: Exploring Emotions in Animals*. New York, Discovery Books.

Burghardt, G. M. (1988) Precocity, play, and the ectotherm–endotherm transition. In Blass, E. M. (ed.) *Handbook of Behavioral Neurobiology*, Vol. 9. New York: Plenum, pp. 107–48.

Burghardt, G. M. (2004) Play: how evolution can explain the most mysterious behavior of all. In: Moya, A. and Font, E. (eds.) *Evolution: From Molecules to Ecosystems*. Oxford: Oxford University Press, pp. 231–246.

Burghardt, G. M. (1999) Conceptions of play and the evolution of animal minds. *Evolution and Cognition*, **5**, 115–23.

Meyer-Holzapfl, M. (1960) Über das Spiel bei Fischen, insbesondere beim Tapirrüsselfisch (*Mormyurs kannume* Forskål). *Zoologische Garten*, **25**, 189–202.

Sherwin, C. M. (2001) Can invertebrates suffer? Or, how robust is argument-by-analogy? *Animal Welfare*, **10**, S103–18.

Hartshorne, C. (1973) *Born to Song: An Interpretation and World Survey of Bird Song.* Bloomington: Indiana University Press.

Lewis, A. C. and Papaj, D. R. (eds.) (1993) *Insect Learning. Ecological and Evolutionary Perspectives.* New York: Chapman & Hall.

Gritsai, O. B., Dubynin, V. A., Pilipenko, V. E. and Petrov, O. P. (2004) Effects of peptide and non-peptide opioids on protective reaction of the cockroach *Periplaneta americana* in the 'hot camera.' *Journal of Evolutionary Biochemistry and Physiology*, **40**, 153–60.

Zabala, N. A. and Gomez, M. A. (1991) Morphine analgesia, tolerance and addiction in the Cricket. *Pharmacology, Biochemistry and Behavior*, **40**, 887–91.

Lozda, M., Romanao, A. and Maldonado, H. (1988) Effect of morphine and naloxone on a defensive response of the crab, chasmagnathus-granulatus. *Pharmacology, Biochemistry and Behavior*, **30**(3), 635–40.

Bergamo, P., Maldonado, H. and Miralto, A. (1992) Opiate effect on the threat display in the crab carcinus-mediterraneus. *Pharmacology, Biochemistry and Behavior*, **42**, 323–6.

Dyakonova, V. E. (2001) Role of opioid peptides in behavior of invertebrates. *Journal of Evolutionary Biochemistry and Physiology*, **37**(4), 335–47.

Panksepp, J. B. and Huber, R. (2004) Ethological analyses of crayfish behavior: a new invertebrate system for measuring the rewarding properties of psychostimulants. *Behavioral Brain Research*, **153**, 171–80.

Ventura, M. U., Montalván, R. and Panizzi, A. R. (2000) Feeding preferences and related types of behavior of *Neomegalotomus parvus*. *Entomologia Experimentalis et Applicata*, **97**, 309–15.

Wäckers, F. L. (1999) Gustatory response by the hymenopteran parasitoid *Cotesia glomerata* to a range of nectar and honeydew sugars. *Journal of Chemical Ecology*, **25**, 2863–77.

Omura, H. and Honda, K. (2003) Feeding responses of adult butterflies, *Nymphalis xanthomelas*, *Kaniska canace* and *Vanessa indica*, to components in tree sap and rotting fruits: synergistic effects of ethanol and acetic acid on sugar responsiveness. *Journal of Insect Physiology*, **49**, 1031–8.

Bernays, E. A. and Bright, K. L. (2005) Distinctive flavours improve foraging efficiency in the polyphagous grasshopper, *Taeniopoda eques*. *Animal Behaviour*, **69**, 463–9.

Chapman, R. F. and Ascoli-Christensen, A. (1999) Sensory coding in the grasshopper (Orthoptera: Acrididae) gustatory system. *Annals of the Entomological Society of America*, **92**, 873–9.

Wheeler, W. M. (1923) *Social Life Among the Insects*. London: Constable.

Eberhard, W. G. (1994) Evidence for widespread courtship during copulation in 131 species of insects and spiders, and implications for cryptic female choice. *Evolution*, **48**, 711–33.

Campbell, V. and Fairbairn, D. J. (2001) Prolonged copulation and the internal dynamics of sperm transfer in the water strider *Aquarius remigis*. *Canadian Journal of Zoology*, **79**, 1801–12.

Vahed, K. (1996) Prolonged copulation in oak bushcrickets (Tettigoniidae: Meconematinae: *Meconema thalassinum* and *M. meridionale*). *Journal of Orthoptera Research*, **5**, 199–204.

Cooper, M. I. and Telford, S. R. (2000) Copulatory sequences and sexual struggles in millipedes. *Journal of Insect Behavior*, **13**, 217–30.

Edvardsson, M. and Arnqvist, G. (2000) Copulatory courtship and cryptic female choice in red flour beetles *Tribolium castaneum*. *Proceedings of the Royal Society of London B*, **267**, 559–63.

Shuker, D., Bateson, N., Breitsprecher, H., O'Donovan, R., Taylor, H., Barnard, C., Behnke, J., Collins, S. and Gilbert, F. (2002) Mating behavior, sexual selection, and copulatory courtship in a promiscuous beetle. *Journal of Insect Behavior*, **15**, 617–31.

Eberhard, W. G. (2002) Physical restraint or stimulation? The function(s) of the modified front legs of male *Archisepsis diversiformis* (Diptera, Sepsidae). *Journal of Insect Behavior*, **15**, 831–50.

Judson, O. (2003) *Dr Tatiana's Sex Advice to All Creation: The Definitive Guide to the Evolutionary Biology of Sex*. London: Vintage, p. 59.

Uhlenbroek, C. (2002) *Talking with Animals*. London: Hodder & Stoughton.

Grove, N. (1996) *Birds of North America*. Hong Kong: Hugh Lauter Levin Associates, Inc.

Pollan, M. (2001) *The Botany of Desire: A Plant's-eye View of the World*. New York: Random House.

Dodson, C. H. (1975) Coevolution of orchids and bees. In Gilbert, L. E. and Raven, P. H. (eds.) *Coevolution of Animals and Plants*. Austin: University of Texas Press, pp. 91–9.

Holldobler, B. and Wilson, E. O. (1990) *The Ants*. Cambridge, MA: Harvard University Press.

Wheeler, W. M. (1910) *Ants, Their Structure, Development and Behavior*. New York and London: Columbia University Press.

Lindauer, M. (1961) *Communication among Social Bees*. Cambridge, MA: Harvard University Press.

Bhattacharjee, Y. (2003) Fly ball or Frisbee, fielder and dog do the same physics. *New York Times*, 7 January.

Fox, D. (2004) Do fruit flies dream of electric bananas? *New Scientist* **181**, 32–5.

Swinderen, B. (2005) The remote roots of consciousness in fruit-fly selective attention? *BioEssays*, **27**, 321–30.

Jackson, R. R., Pollard, S. D. and Cerveira, A. M. (2002a) Opportunistic use of cognitive smokescreens by araneophagic jumping spiders. *Animal Cognition*, **5**, 147–57.

Jackson, R. R., Clark, R. J. and Harland, D. P. (2002b) Behavioral and cognitive influences of kairomes on araneophagic jumping spider. *Behavior*, **139**, 749–75.

Wilcox, R. S. and Jackson, R. R. (1998) Cognitive abilities of araneophagic jumping spiders. In Balda, R. P., Pepperberg, I. and Kamil, A. C. (eds.) *Animal Cognition in Nature*. San Diego: Academic Press, pp. 411–34.

Downer, J. (2002) *Weird Nature*. London: BBC.

Balaban, P. M. and Maksimova, O. A. (1993) Positive and negative brain zones in the snail. *European Journal of Neuroscience*, **5**, 768–74.

Rachlin, H. C. (1976) *Behavior and Learning*. San Francisco: WH Freeman & Co.

Chapter 11

Bridges, R. (1912) *Poetical Works of Robert Bridges*. Oxford University Press: London. (The quoted passage is from 'To Robert Burns – An epistle on instinct,' originally published in 1902.)

Rifkin, J. (2003) Comment: Man and other animals: Our fellow creatures have feelings – so we should give them rights too. *The Guardian*, 16 August.

Cutter, C. (2004) IFAW says whaling hits 15-year high. *US Newswire*. 16 June. [cited December 2005] Available from: http://releases.usnewswire.com/printing.asp?id=31974.

Animal Rights International (2005) *Outlawed in Europe: Three Decades of Progress*. [cited December 2005] Available from: http://www.ari-online.org/pages/europe_3_ summary.html.

European Commission (2001) *Commission Directive 2001/93/EC: Minimum Standards for the Protection of Pigs*. 9 November. [cited December 2005] Available from: http://europa.eu.int/eur-lex/pri/en/oj/dat/2001/l_316/l_31620011201en00360038.pdf.

Davis, K. (2002) The plight of birds in the poultry and egg industry. *United Poultry Concerns*. [cited December 2005] Available from: http://www.upc-online.org/industry/plight.html.

Davis, K. (2005) *The Holocaust & the Henmaid's Tale: A Case for Comparing Atrocities*. New York: Lantern.

Kole, W. J. (2004) *Austria Enacts Strict Animal Rights Laws*. Associated Press, 27 May.

Siebert, C. (2005) Planet of the retired apes. *The New York Times*, 24 July.

Farm Sanctuary (2002) *Florida Passes First U.S. Law Against Cruel Farming System. Sets Nationwide Precedent by Banning 'Gestation Crates.'* 5 November [cited December 2005] Available from: http://www.farmsanctuary.org/media/pr_fl.htm

Humane Society of the United States (2005) *Alternative Energy: Dissection Choice Gaining Ground in Several States*. 14 July [cited December 2005] Available from: http://www.hsus.org/animals_in_research/animals_in_research_news/alternative_energy_dis section_ choice_states.html

McMillan, F. D. (2005) The concept of quality of life in animals. In McMillan, F. D. (ed.) *Mental Health and Well-being in Animals*. Ames, Iowa: Blackwell Publishing, pp. 183–200.

Townsend, P. and Morton, D. B. (1995) Laboratory animal care policies and regulations: United Kingdom. *ILAR Journal*, **37**, 68–74.

de Waal, F. (1982) *Chimpanzee Politics: Power and Sex Among Apes*. Harper & Row, New York, p. 107.

Strum, S. (1987) *Almost Human: A Journey into the World of Baboons*. London: Elm Tree Books.

Brosnan, S. F. and de Waal, F. B. M. (2003) Monkeys reject unequal pay. *Nature*, **425**, 297–9.

Bekoff, M. (2002) Virtuous nature. *New Scientist*, **175**, p. 34.

Pellis, S. (2002) Keeping in touch: play fighting and social knowledge. In: Bekoff, M., Allen, C. and Burghardt, G. M. (eds.) *The Cognitive Animal*. Cambridge, MA: MIT Press.

Masserman, J. H., Wechkin, S. and Terris, W. (1964) 'Altruistic' behavior in rhesus monkeys. *American Journal of Psychiatry*, **121**, 584–5.

Goodall, J. and Bekoff, M. (2002) *The Ten Trusts: What We Must do to Care for the Animals We Love*. San Francisco: HarperCollins.

Beard, M. (2004) The curious incident of the hungry dog in the nighttime. *The Independent*, 5 October, p. 11.

Masson, J. M. and McCarthy, S. (1995) *When Elephants Weep: The Emotional Lives of Animals*. New York: Delacorte.

Griffin, D. R. (1992) *Animal Minds*. Chicago: University of Chicago Press.

Trivers, R. (1971) The evolution of reciprocal altruism. *Quarterly Review of Biology*, **46**, 35–57.

Dugatkin, L. A. (1997) *Cooperation Among Animals*. Oxford: Oxford University Press.

Rilling, J. K., Gutman, D. A., Zeh, T. R., Pagnoni, G., Berns, G. S. and Kitts, C. D. (2002) A neural basis for cooperation. *Neuron*, **36**, 395–405.

Bekoff, M. (2004) Wild justice, cooperation, and fair play: minding manners, being nice, and feeling good. In Sussman, R. and Chapman, A. (eds.) *The Origins and Nature of Sociality*. Chicago: Aldine, pp. 53–79.

Regan, T. (1989) Why child pornography is wrong. In Scarre, G. (ed.) *Children, Parents and Politics*. Cambridge: Cambridge University Press.

Burger, J. (2001) *The Parrot Who Owns Me: The Story of a Relationship*. New York: Villard, p. 128.

Harrison, R. (1964) *Animal Machines*. London: Vincent Stuart Publishers.

Jennings, M., Batchelor, G. R., Brain, P. F., Dick, A., Elliiott, H., Francis, R. J., Hubrecht, R. C., Hurst, J. L., Morton, D. B., Peters, A. G., Raymond, R., Sales, G. D., Sherwin, C. M. and West, C.

(1998) Refining rodent husbandry: the mouse. *Laboratory Animals*, **32**, 233–59.

Merriam-Webster Online Dictionary (2005) [cited July 2005] Available from: http://www.m-w.com/home.htm

EETA/CRABS (Ethologists for the Ethical Treatment of Animals/Citizens for Responsible Animal Behavior Studies). [cited 14 December 2005] Available from: http://www.ethological ethics.org/.

Hagen, K. and Broom, D. M. (2004) Emotional reactions to learning in cattle. *Applied Animal Behavior Science*, **85**, 203–13.

Abeyesinghe, S. M., Nicol, C. J., Hartnell, S. J. and Wathes, C. M. (2005) Can domestic fowl, *Gallus galluls domesticus*, show self-control? *Animal Behaviour*, **70**, 1–11.

Viegas, J. (2005) Study: chickens think about future. *Discovery News*, 14 July [cited July 2005] Available from: http://dsc.discovery.com/news/briefs/20050711/chicken.html

Bekoff, M. and Allen, C. (2004) Cognitive ethology: the comparative study of animal minds. In: Bekoff, M. (ed.) *Encyclopedia of Animal Behavior*. Westport, CT: Greenwood Press, pp. 258–64.

Mason, G. J. (1991) Stereotypies: a critical review. *Animal Behaviour*, **41**, 1015–37.

Garner, J. P. and Mason, G. J. (2002) Evidence for a relationship between cage stereotypies and behavioral disinhibition in laboratory rodents. *Behavioral Brain Research*, **136**, 83–92.

Wemelsfelder, F. (2005) Animal boredom: Understanding the tedium of confined lives. in McMillan, F. D. (ed.) *Mental Health and Well-being in Animals*. Ames: Iowa State University Press, pp. 79–91.

Patterson-Kane, E. G., Hunt, M. and Harper, D. (2002) Rats demand social contact. *Animal Welfare*, **11**, 327–32.

Widowski, T. M. and Duncan, I. J. (2000) Working for a dustbath: are hens increasing pleasure rather than reducing suffering? *Applied Animal Behavior Science*, **68**, 39–53.

Olsson, A. S. and Dahlborn, K. (2002) Improving housing conditions for laboratory mice: a review of 'environmental enrichment.' *Laboratory Animals*, **36**, 243–70.

Sherwin, C. M. (1996) Laboratory mice persist in gaining access to resources: a method of assessing the importance of environmental features. *Applied Animal Behavior Science*, **48**, 203–14.

Krutch, J. W. (1956) *The Great Chain of Life*. Boston: Houghton Mifflin.

Wilson, E. O. (1986) *Biophilia: The Human Bond with Other Species*. Cambridge, MA: Harvard University Press.

Kortlandt, A. (1962) Chimpanzees in the wild. *Scientific American*, **206**, 128–36.

INDEX

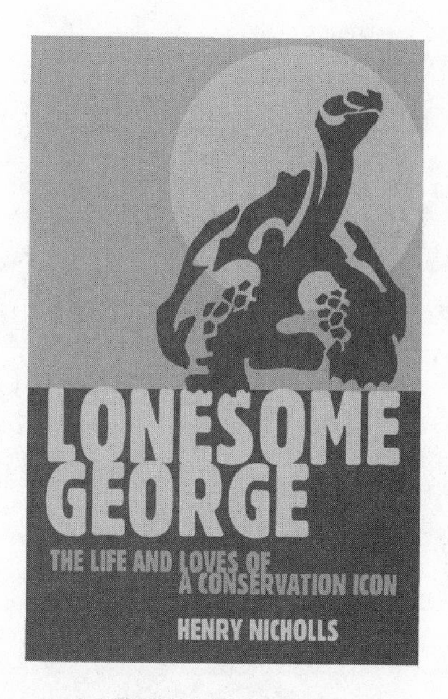
LONESOME
GEORGE
THE LIFE AND LOVES OF
A CONSERVATION ICON
HENRY NICHOLLS